THE USER'S J(
STORYMAPPING PRODUCTS THAT PEOPLE LOVE

Donna Lichaw

Rosenfeld Media
Brooklyn, New York

The User's Journey
Storymapping Products That People Love
By Donna Lichaw

Rosenfeld Media, LLC

457 Third Street, #4R

Brooklyn, New York

11215 USA

On the Web: www.rosenfeldmedia.com

Please send errors to: errata@rosenfeldmedia.com

Publisher: Louis Rosenfeld

Managing Editor: Marta Justak

Illustrations: Eva-Lotta Lamm

Interior Layout Tech: Danielle Foster

Cover Design: The Heads of State

Indexer: Sharon Shock

Proofreader: Sue Boshers

ISBN: 1-933820-31-4

ISBN-13: 978-1-933820-31-6

LCCN: 2015956989

Printed and bound in the United States of America

*For Erica, who begged me to speak and write
so that I would channel my energy
and stop pestering her with my crazy ideas.
I love you.*

HOW TO USE THIS BOOK

This book is meant to be read from cover to cover. It's short and highly scannable, so don't be intimidated. Each chapter builds on the last, introducing concepts and then expanding on how to apply what you've learned to your own practice. This book will also work as a reference after you've read it. The diagrams and illustrations provide you with scannable, short-hand versions of the concepts so that you can go back and jog your memory when needed. So grab a comfy chair and blanket or get comfy on your next flight, sit back, and enjoy. Then keep this book handy for the future because it will likely change the way you work.

Who Should Read This Book?

This book is for anyone who wants to engage an audience by creating things that people want to use, use often, and recommend to others. "Things" can include websites, software, apps, digital or non-digital, for-profit or non-profit services, or even physical goods. I'll collectively call those "things" *products* throughout this book. Whether you are an entrepreneur, designer, product or account manager, content strategist, communications or marketing professional, student, teacher, or engineer, chances are you are someone who can use *story* and its underlying structure and mechanics to make better and more successful products.

What's in This Book?

There are eight chapters in this book grouped into three areas:

- Chapters 1 and 2 discuss how story works and how you can use it to engage your audience not by *telling* stories, but by *creating* stories.

- Chapters 3–5 discuss how story flows through different types of products in different contexts and customer lifecycle journey stages.

- Chapters 6–8 delve into how to uncover, improve, and use your stories both strategically and tactically.

What Comes with This Book?

This book's companion website (📖rosenfeldmedia.com/books/
storymapping) contains a blog and additional content. The book's
diagrams and other illustrations are available under a Creative
Commons license (when possible) for you to download and
include in your own presentations. You can find these on Flickr
at www.flickr.com/photos/rosenfeldmedia/sets/.

FREQUENTLY ASKED QUESTIONS

Is this book about storytelling?

No. And yes. This book is about much more than what you traditionally think of when you think of storytelling. It won't teach you how to *tell* a story. Rather, it will teach you how to *use* story and its underlying structure to craft intended experiences of use that are optimized for audience engagement, similar to what screenwriters and TV writers do with short- and long-form movies and TV shows. Plot point by plot point.

Why story?

We use story because it's one of the oldest and most powerful ways that humans have to communicate with and understand the world. It governs how we do or don't see meaning, value, utility, and affordances in both ideas and things. Story structure and its underlying principles will help you build better products. And it's how you can get your target audience to relate to your product (see Chapters 1 and 2).

Is everything a story?

Yes. Walking down the street? Story. Using an app? Story. Thinking about a product? Story. Using online checking through your boring old bank? Story. Once you start thinking and working like a storymaker, you will ask yourself not *if* something is a story, but if it is or should be a *good* story. The better the story, the more engaged your users will be. Structure is how story engages the human brain (see Chapter 2).

Who is the hero of the stories you map: the business or the user?

As much as you want your business to be the hero of the story, your users are the real heroes. Imagine if *The Wizard of Oz* were about Dorothy, a damsel in distress who is saved by a knight in shining armor. It wouldn't be her story—it would be the knight's story. Dorothy needs to be the hero as much as your customers need to feel like heroes when they find, use, and recommend your product to their friends and family. When you map stories, you're mapping the story you want someone to have with your product. Think of your product as Dorothy's ruby slippers. Without your product, she would never be able to solve her problem. Chapter 2 goes into more detail about how to engineer heroes.

Is storymapping some new process I have to learn?

No. Storymapping is something you can and should seamlessly weave into your existing practice. I want you to start thinking like a storyteller—or story*maker*—so that you can create products that resonate with your target audience. When you start thinking about the story, you'll find that it's the first thing you do at the beginning of any project and something you can easily fold into your existing process. What's the story? You will answer this question by uncovering, mapping, and then testing the story until you get it right (see Chapters 6 and 7).

How do I get started with storymapping?

All you need are some Post-it notes or note cards, a wall or table, some markers, data, and an imagination and understanding of how story works. Once you start seeing stories in your favorite products, you'll see them everywhere. Once you start seeing them everywhere, you'll see how to weave stories into your own work so that you create more successful and engaging products that people love, use often, and recommend to others. Chapters 3, 4, and 5 walk you through how to map different types of stories to solve different types of business and user engagement problems. Chapter 6 tells you how to find stories through research and hypothesis development. Chapter 7 shows you how to use your stories once you've developed them.

What is the difference between *storymapping* and *Agile user story mapping*?

While many people often use the shorthand *storymapping* when referring to Agile user story mapping, they are quite different. Storymapping is as simple as it sounds: literally mapping out an intended experience of use just as you would a story—plot point by plot point. Agile user story mapping is a method that Agile developers use to organize and chart the course for large bodies of work comprised of smaller "user stories (for more on incorporating story development into Agile development, see Chapter 7)." Although the two approaches look similar (Post-its on a wall or cards on a table), they are quite different. Storymapping is a way to engineer increased engagement in your products. Agile user story mapping is a way for engineers to work.

CONTENTS

FOREWORD

I was one of those kids who played *Dungeons and Dragons* (D&D), a fantasy role-playing game that involved going on quests to battle monsters, discover magical items, and drink lots of mead. My friends and I memorized spell books, castle layouts, and Elvish runes, paying more attention to types of armor than we did to types of conjugation for the next English quiz.

In D&D there are two main roles: the player character, who goes on quests in the world of the game; and the Dungeon Master, who operates that world and guides the player characters in their journey.

The first character I played was a wizard. I imagined him being tall, bearded, and wise like Gandalf in *The Lord of the Rings*... but I was young enough that he owed a lot more to Mickey Mouse in *The Sorcerer's Apprentice*. Either way, he was vanquished while trying to cast a sleeping spell on a rather surly bugbear.

Then I tried playing dark, gritty characters like thieves and assassins—they talked tough and fought tougher. Why, my 7th-level Rogue wouldn't give a bugbear the time of day! Even so, he met his fate while pick-pocketing an unusually large stone giant who, drunken on mead, sat on him.

When I became a Dungeon Master, I didn't want the players in my game to die as quickly as mine always did. Rather than crushing my players' dreams with an overly hostile world, I wanted them to have a chance of reaching their goals. That would make for a more interesting game and a far better story. But with my limited experience (I was 13), I didn't know how to start telling the tale.

You may feel the same way. If you build products or design services, you know how easy it is to get ambushed by constraints, surprised by your competition, and buried in strategies dark and deep. You may find that coming up with that next iteration is much harder than you thought, or you may get usability feedback that changes your entire approach. You may face indecision or conflicts on your team that keep you from moving forward. And even when you're armed with data and research, it can often seem like you're on a quest with an uncertain ending.

Donna Lichaw is here to help. Drawing on her experiences with Fortune 500 companies, public radio, filmmaking, and more, Donna helps you navigate the oft-treacherous waters of product development. She helps you not just to *tell* stories or *use* stories to promote your product, but to build your products *as if they were stories themselves*.

Why stories? Because they're our oldest, best tools for communicating, teaching, and engaging with people. Because they help us understand the landscape of how people interact with our products. And because they help us understand the people themselves.

Using Donna's approach, you'll cast your users as the heroes of the story so that everything you do supports them in their journey. And when you help your heroes overcome their challenges, surpass their obstacles, and make progress toward their goals, you'll also take steps toward your own.

Like Donna says, "I wish it were more complicated, but it really is that simple."

So ready your armor, grab a cup of mead, and roll the dice. Here there be dragons, but fear not—Donna gives you the key to defeating them: *story first*.

—Jonathon Colman
Product UX + Content Strategy, Facebook

Note: All content and viewpoints expressed here solely reflect the thoughts and opinions of the author.

INTRODUCTION

"How do you build your storyline? By using 3 × 5 cards."

—Syd Field,
Screenplay

I n his classic tome on screenwriting, Syd Field claimed that he could not teach aspiring filmmakers how to write a screenplay. "This is not a 'how-to' book," he explained. "People teach themselves the craft of screenwriting. All I can do is show them what they have to do to write a successful screenplay. So, I call this a *what-to* book..."

What Would MacGyver Do?

MacGyver, the eponymous star of the 1980s television show of the same name, could solve any problem or get out of any situation with a needle, some thread, and bubble gum.

Storymapping is much the same. If MacGyver built products, he would map stories. Storymapping can help you solve any engagement-related problem with your product or even create a successful product by mapping the story before you design or build anything.

How do you map a story for your product? All you need are some Post-it notes or note cards, a marker or pen, a whiteboard or wall, data or an imagination, and an understanding of how story works. Then you map your story. Plot point by plot point. There is some trial-and-error involved at first, but once you build your story muscle, you'll be storymapping like a champ.

I wish it were more complicated, but it really is that simple. And fast. You can do it alone, but I recommend doing it with a team for maximum efficiency and buy-in. While I can't tell you much more in the *how* department, I can show you *what* it takes to build a successful story that works—for you, your customers, your product, and your business. I can also show you how to apply stories once you've created them and give you some rules of thumb to set you on the right path.

Let's say that you want to build a new product, but aren't sure if it's a good idea? That's a story. You want to help people find your product?

Also a story. You want to get people to try your product out? Yup, story. You want to figure out how your product should work? Story. People try your product, but don't return to use it again? That's a story, too. A cliffhanger of a story and one that you can easily fix with some props and ingenuity. Just like MacGyver.

You'll learn how to ask three simple questions before you start any new project:

- What's the story?

- Who is the hero?

- What is the hero's goal?

After a while, you won't just be asking what the story is, but whether it's a *good* story. Because a good story isn't just a random series of events—that's a flow chart or a terrible student film. A good story makes things go *boom!* For your customers. And for your business.

Because Structure Is Key

The book is split into three parts. In the first part (Chapters 1–2), you'll learn why story matters for things that aren't just entertainment, fiction, or movies, as well as how story functions in products and services. In the second part (Chapters 3–5), you'll learn about different types of stories and how those frameworks flow through successful products. Finally, in Chapters 6–8, you'll see how to apply stories to your own work, in different contexts, so that you can build successful products that resonate with your target audience. By the end of this book, you'll think like a story*teller* and work like a story*maker*.

Mapping
the Story

"You need a road map, a guide, a direction—a line of development leading from beginning to end. You need a story line. If you don't have one, you're in trouble."

—Syd Field,
Screenplay: The Foundations of Screenwriting

In 2004, I presented my year-end documentary film in graduate school to an audience of around 100 people. As soon as the film ended, before the lights went up, one of my classmate's hands shot up. I will never forget the first words he uttered—they're etched into my brain.

I can't believe you made me sit through that. What was the point?

My film was a dud. It had nothing holding it together: no conflict, no climax, and no resolution—ergo, no story. As a result, I failed to engage my audience. I somehow forgot one of the foundational tenets of filmmaking: if you want to engage your audience, your film *must* have a story at its foundation.

A website, software, app, service, or campaign—for brevity's-sake I'll use the term *product* for the rest of the book—is similar to a film. They are all things that humans *experience*. Just like with a film, if you want to engage your audience, your product must have a story at its foundation. You can do this by accident like I did when I created films that people loved. (I did have a few of those, I promise.) Or you can map the story with deliberate care and intent like I eventually learned to do, both as a filmmaker and more recently as someone who helps businesses build products that people love.

Making Things Go BOOM!

Vince Gilligan, creator of the television show *Breaking Bad,* knows a thing or two about using story to engage an audience. In this photo (see Figure 1.1), he is seated in front of the story map for Season 4.

TV writers are smart. They map the story out *before* they write a line of dialogue or shoot a single scene. TV shows are large, complex things that are built with large, distributed teams over a long period of time. With so many people, scenes, episodes, and seasons to manage, it's hard to stay focused on the big picture. Mapping the story on a wall helps TV writers plot a course while keeping the big picture in mind.

FIGURE 1.1
Vince Gilligan, creator
of *Breaking Bad*, in
front of a story map
for season 4.

Mapping the story also helps TV writers build a product that engages an audience by adhering to a millennia-old architecture designed for engagement: a well-crafted story. You'll learn more about story architecture in Chapter 2, but in the meantime, consider this meticulously placed card near the end of the storyline for *Breaking Bad,* Season 4 (see Figure 1.2). This card has one word written on it: "BOOM." If you've seen Season 4 of *Breaking Bad*, you know what this refers to. If you haven't, you can imagine. Mapping the story helps TV writers make things go BOOM. And it will help you, someone who builds products, make things go BOOM as well.

Story is why people tune in and stayed tuned in, whether you're creating a TV show, a movie, or a website. Storymapping is how you make that story happen, whether you're a screenwriter or a product person.

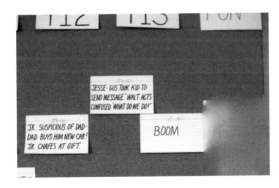

FIGURE 1.2
A close-up of a story
card for *Breaking Bad*.

Storymapping is just what it sounds like: mapping out an intended experience of use for a product, plot point by plot point. This concept of mapping stories is not new. It's something that Aristotle started doing a very long time ago as he sought to understand what it was about Greek dramas that enraptured audiences, so their success could be reproduced. It's something screenwriters have been doing for years. It's something I eventually learned to do for films and more recently with products. And it's something that you can do on your next project or product.

Why Story?

Story is one of the most powerful tools that humans use to understand and communicate with the outside world. Part evolutionary feature, part survival mechanism harking back to Paleolithic times, part communication tool—story powers the human brain. Story-based cognitive function is so powerful that neuroscientists have a term for it when it doesn't work: *dysnarrativia*, the inability to understand or construct stories. Narrative cognition is so central to how humans operate that not having it is debilitating. Like living with amnesia, it is difficult, if not impossible, to function normally. Story, and its underlying architecture, powers the ability to understand what happened in the past, what happens in the moment, or what will happen in the future. It's a framework and a lens with which humans comprehend everything.

Whether you plan for it or not, your customers use their story-driven brains to understand your product and what it's like to use your product. They also use their story-driven brains to tell others about your product. The better the story, the better the experience, the better the word of mouth.

More specifically, when people experience something with a story at its foundation—whether it entails watching a movie, riding a roller-coaster, or using a website—their brains are activated. They are more likely not just to have a *good* experience, but to:

- Remember the experience.
- See value in what was experienced.
- See utility in what they did during that experience.

- Have an easier time doing whatever they were trying to accomplish.

- Want to repeat that experience.

All of this fits under the umbrella of *engagement*.

If you're in the business of building products that engage, it's your job to consider the story that you and your business want your customers to experience. In this book, you will learn how to map that story—or stories—and align everything you and your business do so that it supports that story. For your customer. And your business.

It works for movies, and it will work for you.

NOTE *THE* OR *A* STORY VS. STORY

It may look as if I've made a mistake throughout this book by using the word "story" without an article in front of it, i.e., *the* story. It's no mistake. Story is as much of a tool and framework as it is a discipline. Like art. Or science. When I use "story" without the article, I'm talking about story as a tool. For example, I might say "use story to turn data into insights." However, if I refer to "a story" or "the story," I'm referring to the thing you will create and weave throughout your work.

How Story Works

"For, the more we look at the story (the story that is a story, mind), the more we disentangle it from the finer growths that it supports, the less shall we find to admire. It runs like a backbone—or may I say a tape-worm, for its beginning and end are arbitrary."

—E. M. Forster,
Aspects of the Novel

Humans are sense-making creatures, and story is our most critical sense-making tool. As humans, we've evolved and innovated story over millennia as a way to understand our world. For example, there is evidence that ancient cave dwellers learned how to trap an animal and not go hunting alone through the use of stories.

Given how long we've lived with story, it's not surprising that Aristotle uncovered a working model for it long ago. Basically, he said that every story needs three things: characters, goals, and conflict. What weaves these elements together is a structure or a series of actions and events that have a shape to them.

Fortunately, story and its underlying structure is straightforward, simple, and can be easily learned. That's why it's so powerful—for books, films, and products.

NOTE STORY VS. NARRATIVE VS. PLOT

Don't think that you're alone if you confuse the words story, storyline, plot, and narrative as they are often used interchangeably. Even the dictionary will define them similarly. While story or narrative can refer to the broader version of events, and plot or storyline breaks down the "plan" or series of actions and events that lead up to the story, I use them in this book interchangeably. The reason for this is that a story without a plot or storyline (i.e., without a structure) is just a random series of events. Random series of events don't make for good stories and definitely won't engage your target audience. Your story must engage your audience. And in order for it to do so, it must have a structure at its foundation.

Story Has a Structure

First, every story has a beginning, middle, and end—with the middle typically taking up a longer period of time than the beginning or end, as shown in Figure 2.1.

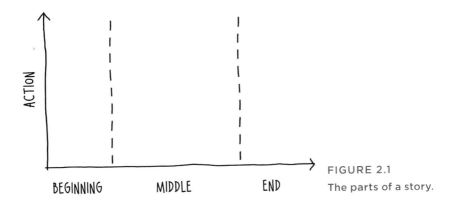

FIGURE 2.1

The parts of a story.

Next, every story has a structure, similar to what you see in Figure 2.2. It's typically called the *narrative arc* or *story arc,* which is a chronological series of events.

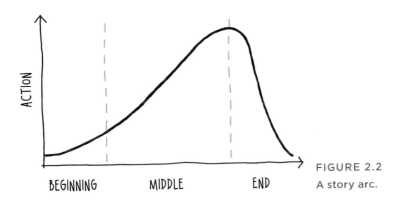

FIGURE 2.2

A story arc.

While the X-axis in Figure 2.2 represents time, the Y-axis represents the action. In other words, you can visually see in the figure that the story builds in excitement, the pace of its action increases over time until it hits a high point, and the story winds down before it ends. When the story doesn't wind down and instead ends while the action is still rising or at a peak, the story is called a *cliffhanger.*

Every narrative arc has specific key plot points and sequences, as shown in Figure 2.3.

Let's dissect the narrative arc of a story. Narrative arcs are comprised of the following elements:

- Exposition
- Inciting incident or problem
- Rising action
- Crisis
- Climax or resolution
- Falling action or denouement
- End

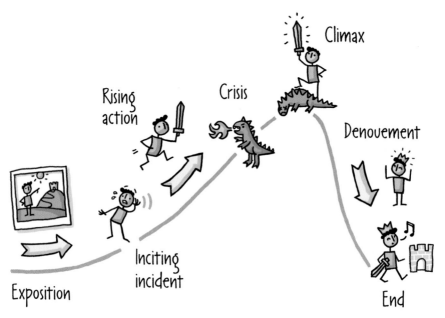

FIGURE 2.3
Plot points on a story arc.

Exposition

During the exposition, you are introduced to the world of the story, the characters, and some kind of big goal. There is a main character, and that character wants something. Big. The exposition functions not only to set the stage of a story, but also to get the person on the other end—*you*, the viewer—interested and engaged with the main character or characters and what drives them. At its most powerful, a good exposition will compel you to see yourself in and identify with a character or a set of characters. At the very least, it compels you to empathize with them.

Take for example the 1985 feature film, *Back to the Future.* In the exposition, you meet Marty McFly, who lives in Hill Valley—any suburb USA. His family isn't very ambitious, but Marty has plans. He's going to make something of himself. Marty has a friend, Doc—a mad scientist who built a time machine (see Figure 2.4). All very cool.

FIGURE 2.4
In the exposition to
Back to the Future,
you are introduced to
Marty, Doc, and their
time machine.

In a movie like *Back to the Future*, you're compelled to empathize with Marty. You don't have to *like* him. You just have to understand him, his goals and why he wants to pursue them.

Inciting Incident or Problem

The inciting incident is the moment where something changes or goes dramatically wrong in the world of the story. A problem surfaces and gets in the way of the character meeting his big goal. The moment when the hero is thrust into leaving his safe world in order to fix the problem is called the *call to action.*

Neuroscientists have shown that when you listen to or watch a story, it's as if you are experiencing the story in real time. As action rises, your pulse might quicken or your palms get sweaty. Something startles you, and you jump. Stories are not just about looking or listening, they are about *being*. The inciting incident is the first hook or trigger point in a story that amplifies *if* and *how* you identify with the main character, what problems he has, and what he has to go through to fix that problem and meet his goal. It's what gets you hooked. When the main character is called to action, it's as if you, the viewer, are called to action. Your brain starts working in overdrive to figure out what will happen next and how the hero will right the wrong.

In *Back to the Future*, the excitement of a time machine doesn't last long; militants shoot Doc in a parking lot in an attempt to retrieve plutonium that he stole from them (see Figure 2.5). Not good. In an attempt to escape, Marty ends up driving the time machine to 1955 and then finds out that he can't get home. That's a problem—a meaty enough problem to name the movie after. Marty's call to action is simple: to get back to the future.

FIGURE 2.5
Inciting incident: this is the point in *Back to the Future* where the story kickstarts into action.

Rising Action

After the problem surfaces in the inciting incident, the protagonist of the story goes on a journey to right that wrong. We spend the rest of the story not just *seeing* how it all pans out, but also *feeling* how it pans out. A good story escalates during the rising action, creating new tensions and conflicts that help move the story forward. As the story builds, the audience's anticipation and excitement builds simultaneously.

During the rising action of any good story, there is also plenty of conflict to keep the audience engaged. Without conflict, endings come too easily, and the audience is unconvinced or bored or both.

In this sense, humans are easy—because to keep them engaged, you save the best for last.

In *Back to the Future*, Marty sets out to find 1955 Doc. They try to get Marty home. But they can't solve the problem, or the movie would end. So things get weird. Marty meets the younger version of his mom. His mom has a crush on him (see Figure 2.6). Marty meets his dad's nemesis, Biff. Biff becomes Marty's nemesis. Things get tense. And more tense. And as a result, more engaging. We become more and more invested in how Marty will get back to the future, scene by scene.

FIGURE 2.6
In this scene, Marty starts to realize that his mom is taking an interest in him and his Calvin Kleins.

Crisis

The story culminates at the point (or series of points) of maximum conflict—the crisis. It's the point of no return. Nothing the hero has done has worked, and he is further from the goal than ever. The story either has to right itself or end in tragedy. If the story ends with neither, then it's a cliffhanger and is incomplete. At this point, the main character has gotten so far and is so close to meeting his goal that it's impossible to just give up. Defeat or success is the only option.

In *Back to the Future*, the crisis starts when Marty is close to figuring out how to get back to the future. But because his mom falls in love with him instead of falling for his father, there is the chance that he will never be born in the future. Because he might never be born, Marty begins to disappear (see Figure 2.7). The only way to get over this hurdle is to make sure that his parents end up together. But how? And then what? Once he overcomes this obstacle, he still has to figure out how to get home. How will this all play out, you wonder, as you are now totally invested in the outcome of the story.

FIGURE 2.7
Marty starts to disappear while he's on stage at a school dance. Will he or won't he get his parents together so that he can live? Suspense!

Climax or Resolution

Just as it sounds, the climax occurs at the top of the story arc. It's the most important part of the story. It's the high point. The final show-down. This is the point at which the hero's fate and the direction of the story are determined. As such, it is also the most exciting part of the story. It's the point at which all of that tension and *will he or won't he* from earlier scenes culminates in you jumping out of your seat, cheering, laughing, feeling satisfied that you solved the mystery before the main character, or simply smiling because well…that was awesome.

Climax is why you are glad that you tuned in and stayed tuned in.

Sometimes, this point is also called the *resolution*, which occurs when the main problem from the inciting incident and the hurdle from the crisis are resolved. Problems and hurdles are either resolved or they're not, and you're left with that tragedy or cliffhanger.

In *Back to the Future*, the climax begins when Marty's parents kiss at the high school dance. At the very least, this means that he can finish playing his song on the guitar. Excellent.

But wait!

There's more!

There's a clock tower and lightning (see Figure 2.8). The underlying problem still needs to be solved: Marty needs to figure out how to get home. In a bolt of lightning, boom, Marty gets catapulted back to the future. Even more excellent.

FIGURE 2.8
Climax and resolution in *Back to the Future.* Doc figures out how to get Marty home by harnessing lightning to power the time machine from atop a clock tower.

Think of the climax as a sort of pay-off. This is why you sat through one hour and 30 minutes. It's exciting. It's suspenseful. It's satisfying. The climax is not only the best part of the story, but it's what you remember most. It's why you come back. It's like a reward, or a thank you for tuning in and staying tuned in.

But then what? Imagine if Doc managed to harness the power of lightning, get Marty home, and the movie just ended. Stories can't just end on a high point, or they're as unsatisfying as a cliffhanger. Once Marty gets back to the future, he still needs to actually get home. For that, you have the falling action or denouement.

Falling Action or Denouement

Have you ever listened in frustration to someone having a conversation on her mobile phone? If you had to listen to the entire conversation in person, with both participants audible, it wouldn't be nearly as frustrating. You could probably tune the conversation out or listen and just not care. It turns out that the main reason why these conversations are so frustrating is that your brain naturally wants to complete the conversation. Just hearing half triggers an automatic, unsatisfying response that leads to frustration. Researchers call this phenomenon a *halfalogue*: half of a conversation that your brain naturally and uncontrollably tries to complete.[1]

Humans, it turns out, need closure. Stories, likewise, need closure so that humans can *feel* closure.

Imagine if *The Wizard of Oz* ended after Dorothy had defeated the wicked witch. Goal met. The end. You'd be frustrated. Your brain would jump into overdrive as you wondered *what then? What about Kansas?* Your mind would jump full circle as you started to remember the exposition of the story and wanted to know not just how evil was defeated, but how the story *ended*. What happened to Dorothy after she defeated the witch? For this reason, stories need not just to resolve their conflict and show characters meeting their goals, but also to have a fancy ending called a *denouement*, a word derived from the French meaning "to unknot." This is the part of the story when the conflict is resolved and the action starts slowing down in pace and excitement toward the closing scene. It's how everything in the story gets wrapped up.

The line between the climax/resolution, falling action, and ending can be blurry and happen so quickly that it's hard to discern the difference between one and another. What matters is that the climax is exciting, and it resolves the major conflict or problem—the falling action leads to closure. In *Back to the Future*, Marty McFly goes home (see Figure 2.9). This is the falling action for many adventure tales: the hero goes home.

Tension releases. Ah…all is good in the world.

And…it's important that home is better than when the story started and where the character left it. In this case, it's *much* better. Marty's parents are successful. Biff is his family's servant. Marty got the truck he always wanted. Not bad.

Ideally, the falling action or denouement should happen as quickly as possible. As much as humans need closure, they're also impatient beings. Once the action has died down, there is only so much that can keep your attention. Just because you want closure, doesn't mean it needs to be dragged out with a 10-minute long ticker-tape parade. (I'm looking at you, George Lucas.)

---

1 http://www.news.cornell.edu/stories/2010/05/half-heard-phone-conversations
   -reduce-performance

End

Quite literally, the end *is* the end. Characters grow throughout a story and should be changed by the end. Remember that big goal established in the exposition? How did it all work out? At this point, the character should meet her goal and hopefully learn something along the way. Along the same lines, just like your cave-dwelling ancestors, *you* should be changed and have had a new experience, or have learned something by the end of a good story.

In *Back to the Future*, the story ends with Marty's girlfriend asking him if everything is OK. "Everything," Marty says, "is perfect." They embrace (see Figure 2.10). Now, if this were a classic Hollywood film, the two would kiss, the screen would fade to black, and the credits would roll. The end.

FIGURE 2.10
All is well. Marty and his girlfriend embrace.

But as you may remember, this is the first installment of what would become a trilogy. Before you get too comfortable in your plush movie seat or sofa, you see a flash and Doc running up the driveway. Something's not right. There's a problem and Doc needs help. Where does he want to take them? To the future! And so a new story is kickstarted…a sequel. Just because a story has comes to an end and has closure doesn't mean it can't lead to another story…and another. We call those *serial stories*. Serials keep us engaged episode by episode. Serials are fun.

Building Products with Story

> "...in real life we each of us regard ourselves as the main character, the protagonist, the big cheese; the camera is on us, baby."
>
> —Stephen King,
> *On Writing: A Memoir of the Craft*

Let's face it: you probably don't make multi-million dollar epic movies like Star Wars; instead, you make websites, software, digital or non-digital services—all things that people don't just consume, but actually use. Just as with a movie, story flows through how people find, think about, use, and recommend your products.

Consider this photo for a moment (see Figure 2.11). It tells a story of an Apple product that comes installed on every iPhone. You can probably guess what product it is.

FIGURE 2.11
A still image from an Apple commercial showing two people using a built-in iPhone app.

Assuming you guessed FaceTime, you are correct. If you guessed "Tinder for seniors," that's not an Apple product. But, as some of my past workshop attendees have demonstrated, a product like that also has a compelling story to it: a story that you can easily use to prototype to test out a design hypothesis. What we see in this still photograph is an entire story encapsulated in one simple frame. Rather than spell it out for you, I want you to take a moment to consider this narrative within the framework I've laid out so far.

What do you see?

How do you know that these people are using FaceTime?

Well, they're older, so maybe they're grandparents. They're smiling. What makes grandparents smile? Grandchildren? And? Maybe their grandchildren are far away, and they want to see them. Why can't they see them? It's too expensive to fly and not realistic to do that on a regular basis. Why not call them up? They already have an iPhone or an iPad and use it to play the crossword puzzles all day. And so forth…they *are* calling them up. Just with video. Using FaceTime is as easy as using the phone. It *is* a phone. But with video. You just look at it instead of holding it up to your ear…like magic.

This is the type of computational math that your brain makes during a series of microseconds when you look at a photograph like this and try to understand what you see. Your brain seeks out a story in the data it consumes. And that story has a structure to it, whether you realize it or not. This behavior is so natural that you probably don't even notice that you do it.

Story is not only a tool your brain uses to understand what you *see*, it's a tool your brain uses to understand what you *experience*. In other words, the same brain function that you use to understand what you *see* in a photograph is the same brain function you would use if you were one of those grandparents *using* FaceTime. Life is a story. And in that story, you are the hero.

In *Badass: Making Users Awesome*, Kathy Sierra argues that creating successful products is not about what features you build—it's about how *badass* you make your user on the other end feel. It's not about what your *product* can do, but instead about what your *users* can do if they use your product.

Amazon, for example, is not a marketplace with lots of stuff. It's a way for you to have a world of goods at your fingertips. Using this perspective, you can see how your job building products comes down to creating *heroes*. When I rush-order toothpaste with one-click on Amazon to replace the toothpaste I used up this morning—as boring as it sounds, I'm a hero in my household.

This job you have of creating heroes isn't just an act of goodwill. In the time I've spent over the past two decades helping businesses build products that people love, I've seen what happens when people feel good about what they can do with your product. They love your product. And your brand. They recommend it to others. They continue to use it over time…as long as you keep making them feel awesome. They even forgive mistakes and quirks when your product doesn't work as expected, or your brand doesn't behave as they'd like. People don't care about your product or brand. They care about themselves. That's something that you can and should embrace when you build products.

What's great about story and its underlying structure is that it provides you with a framework—a formula, if you will—for turning your customers into heroes. Plot points, high points, and all. Story is one of the oldest and most powerful tools you have to create heroes. And as I've seen and will show you in this book, what works for books and movies will work for your customers, too.

CHAPTER 3

Concept Stories

"Stories are about people, not things."

—Chris Crawford,
Chris Crawford on Interactive Storytelling

When the first iPhone came out in 2007, the iPod was a popular device. If you were like me, you carried an iPod in one pocket and a mobile phone in another. Sometimes, you joked about how you wished you could duct tape them together so they could be one device. But really, you wanted Apple to invent an iPod that was also a mobile phone.

When Steve Jobs gave his keynote presentation in January of 2007, that is exactly what the media and pundits expected him to announce. And he did announce an iPod that made phone calls. Sort of. What he demonstrated to the world in that presentation surprised people because it was much, much more.

During his keynote presentation, Jobs presented a problem: smartphones are no good. Then he revealed a new smartphone that not many people expected—it consisted of not one, but *three* products:

- A widescreen touchscreen iPod

- A revolutionary new mobile phone

- An Internet communicator

As he cycled through three slides in his presentation that illustrated these three points, he repeated them a few times. "An iPod...a phone... and an Internet communicator..." he repeated this phrase until he finally asked the audience, "Are you getting it?" At this point, the audience erupted in applause as he announced that Apple was not launching three, but one singular device that did all three things. They were going to call it the *iPhone.*

No one had asked for a three-in-one communication device. Actually, most iPod owners in 2007 would have been content with an iPod that let them make phone calls. This moment in Apple's keynote presentation was not just momentous because it changed the world of mobile computing, but also because it was the inciting incident that kick-started a storyline that flowed through everything from the actual product itself to the rest of the presentation that hooked and engaged not just the audience, but much of the world. What bolstered the presentation, more specifically, was a *concept story.*

What Is a Concept Story?

A concept story is the conceptual story model of your product: it illustrates the big picture overview of what a product is. At the highest level, it also outlines how your customers *think* about that product. It is the foundational story and structure that you will use to identify and communicate your core concept and value proposition both internally and externally, as well as weave into everything you eventually build.

Concept stories, when used to define products, help you answer the following questions:

- Who is this product for?
    - What is their problem?
    - What is their big goal? Secondary goals?
- What *is* this product?
- What is the competition?
- Why might someone *not* want to use this product?
- How is this product better than the competition?
- What does this product need to do?
    - What is the straightforward solution to the problem?
    - What is the *awesome* solution to the problem?

> **NOTE** WHAT CONCEPT STORIES DO
>
> At the very least, good concept stories get people excited about your product. As a requirement, the stories live within your product and how you shape it. At their best, they get people *talking* about your product. Concept stories help you achieve three goals:
>
> - Communicate a shared vision
> - Align toward that shared vision
> - Innovate and prioritize against that shared vision

How Concept Stories Work

Because concept stories illustrate how your target customers do or could think about your product or service, they are either based on real data or are aspirational. Think of them as the mental calculation that someone makes when they first hear about your product. The story might only last a few seconds as your customer puts together the important pieces of what your product is and what they can do with the product. Even though it lasts a few seconds, this story sets the stage for your customer being intrigued or excited by what your product is.

Concept stories operate like this (see Figure 3.1):

- **Exposition:** The current state of things

- **Inciting Incident/Problem:** The problem your product will solve

- **Rising Action:** The product name and a brief description or market category

- **Crisis:** The competition

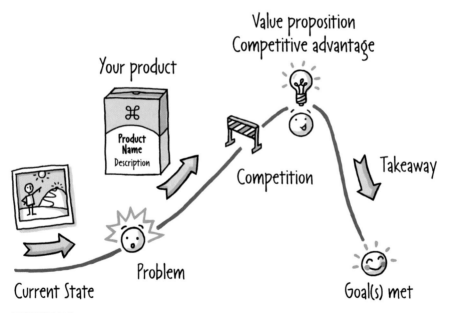

FIGURE 3.1

How a concept story is structured and operates—this is how people think about and see value in your product.

- **Climax/Resolution:** The solution and value proposition or competitive advantage

- **Falling Action:** The takeaway

- **End:** The goal met

Exposition

Exposition reflects the current state of things for your user who personifies your target audience (see Figure 3.2). Who is that user? What does he want? What does she need to do?

In the case of the first iPhone, the story exposition began with a character who loved her iPod and her mobile phone, but wanted a device that would let her listen to music and make phone calls on the go. If you asked why she wanted those things, you'd see that both of these things fall under the umbrella of communication: a basic necessity. Your character doesn't want a two-in-one device, per se, but just needs to communicate with the world.

FIGURE 3.2
Identify your main character or user.

While writing this book, I struggled with what to call the main character. This "person" can go by many names: person (obviously), character, hero, user, customer, target audience, persona, etc. In the end, I settled for using the words *character* or *user*, as those two names seemed most apropos. A *character* is typically characterized in a story, and a *user* typically represents the business customer. If I used other words occasionally, they are intended to mean the same thing.

Inciting Incident/Problem

The inciting incident is the problem or need that your users have. They have a big goal, but...wait...there's a problem. Why can't they meet their goal?

If there isn't a problem, then there is no solution...and without either, there is no story. The problem doesn't have to be very serious or a matter of life and death. It can be as simple as boredom. This problem might be one the users know they have or one that you need to show them they have. Both are valid. Additionally, this is a problem that they can likely solve through other means. Rarely will you be inventing a product that is exploring completely uncharted territory. Even the iPhone was solving a problem that other competitors were trying to figure out: it's difficult to communicate while on the go.

In the case of the iPhone, the problem that the user knew she had was that it sucked to carry two devices. The problem that the iPhone ultimately solved, however, was more broadly focused on improving mobile communication. In this case, Apple solved a problem that people didn't know they had. As such, the 2007 keynote, as well as the device itself, not only had to *tell* the world what their problem was, but also *show* what the problem was and *highlight* how the solution could look and function.

Rising Action

The rising action occurs when your product, service, or feature comes to the rescue. The product should have a name, a brief description, or a market category (see Figure 3.3). For example, the iPhone is a smartphone, specifically, and a mobile communication device, more broadly. Because concept stories are short and conceptual in nature, the rising action shouldn't be too complicated or wordy.

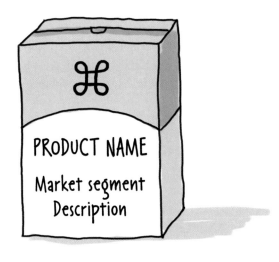

FIGURE 3.3
Give your product an
identity as well.

> **NOTE** CONCEPT STORY: A VISUAL ELEVATOR PITCH
>
> Think of a concept story as a way to visualize and bolster
> a short, impactful, bulletproof elevator pitch. Both concept
> stories and pitches describe your product, brand, or business,
> as well as purpose, market, value propositions, competition and
> competitive advantage (more on elevator pitches in Chapter 7,
> "Using Your Story").

Crisis

Think of the crisis as the competition. This competition can be
another product, service, or feature. It can be abstract, as in an
alternative way that people currently solve their problems or meet
their needs. Or it can also be something emotional, such as resistance
to change or people not wanting to adopt something new.

In the case of the first iPhone, the crisis involved a little bit of all of
the above. Users might already own an iPod, mobile phone, or both
and not want to buy a new device. If they were interested in buying a
new device, however, they might not want to pay a lot for it. If they'd
ever used a touchscreen device before, knowing that the iPhone
featured a touchscreen was also a huge crisis: touchscreens were as
awful as the smartphones they accompanied those days. Wouldn't
a touchscreen make the iPhone difficult to use? And no keyboard?
Typing would be impossible. At least that's what the few people who
owned Blackberrys and Palm Treos thought at the time.

Climax/Resolution

The climax is where the problem outlined in the inciting incident and the additional hurdles that surface during your crisis are resolved and overcome. The way that your product enables users to resolve these problems becomes its value proposition. Implicit in the value proposition is that it's not only different, but also better than the alternative ways your customer has to solve this problem. A concept without a conflict and a resulting climax is a flat story—literally just a line.

In the case of the iPhone, if the character wanted a two-in-one device to communicate, what they get with the iPhone was the *best* way to communicate. With the iPhone, not only could they listen to music and make calls, but they could also access the Internet, maps, and email.

Granted, other smartphones would let them do some of these tasks, but as Steve Jobs emphasized over and over during his keynote presentation, the iPhone worked like magic. It was easy to use. Those simple capabilities and value propositions fit neatly on a business school competitive advantage graph, like the one Jobs mentioned in his keynote. And they gave a strong climax to his story at the conceptual level. Who doesn't want a bit of magic in their lives?

Falling Action

The falling action is the part of the story where your hero has some kind of takeaway—when he envisions a path to try out, use, or purchase the product. Think of this as the *then what?* or *...and?* Your product solves a problem and overcomes the competition in a compelling way. So what? If a product falls in a forest and no one hears it, what's the point? Use the falling action of your concept story as a chance to empathize with your character and imagine how you want that person to think or feel. Is what you want to happen plausible? If so, how? If not, why?

For the iPhone, the falling action for the character at the center of this story was that either she was convinced that she wanted this device (early adopters, fanboys, and fangirls), was intrigued and needed more convincing to try it out (she might wait and buy the iPhone 2), or she stayed skeptical but curious (your grandparents waited a long time and eventually bought an iPhone). Falling action for concept stories should still remain in the realm of thought, rather than action. Your customer hears about what she can do with your product and *thinks* something. In the next chapter, you'll see how you can move that person from *thinking* to *doing*.

Without this falling action, the story is either not complete enough or the target market isn't right for the product. As with all plot points, if you use real data to build your story, it will be that much more powerful. If you posit this falling action as a hypothesis, you should test it with real people. For example, if you think that people would want this device, but aren't sure, test your story out by talking to or surveying your customers or testing a prototype in the wild (more on story validation in Chapter 7).

> **NOTE** DIFFERENT STROKES FOR DIFFERENT PERSONAS?
>
> If you find that the concept stories for different personas are drastically different from one another, you can draft them each separately. Likely, you'll be able to draft them together and note the differences. Those differences will come in handy as you move into drafting origin and usage stories (discussed in Chapters 4, "Origin Stories," and 5, "Usage Stories").

End

Simply put, at the end, your customers can see themselves meeting their goal. At this point, your high-level business goal or mission must also resonate with the story. For example, if your business's mission is to help people find love, and your user's goal is to find love, great. The story works for both your user and the business. Even if your business's goal is to sell ad space, and your user's goal is to learn something, your story works for both. All of the plot points that lead to this moment make sure that it all comes together.

With the iPhone, the character knows that she can get her music and phone all in one place *and* communicate with the world around her (see Figure 3.4). And at the very highest level, Apple helps people communicate better. This mission is lofty, which is good. In the following chapters, we'll discuss how to get people to start using your product.

FIGURE 3.4
The concept story model for the first iPhone.

Avoiding the Anticlimactic

Six months before the iPhone 1 announcement, Apple filed a patent for a different kind of device—one that looked and would function drastically differently than what they would eventually launch.[1] The patent drawings for what could have been the first iPhone were a logical solution to a known problem. The device was essentially an iPod that made phone calls (see Figure 3.5).

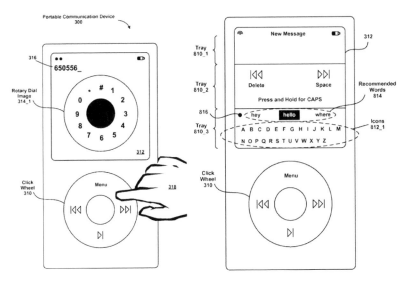

FIGURE 3.5

A diagram from Apple's patent application for what could have been the first iPhone.

If you lay it out like a story, you can see that the prototype illustrated in the patent specifications is *anticlimactic* (see Figure 3.6). Literally, it lacks a climax:

- **Exposition:** I love my iPod; I love my mobile phone.

- **Inciting Incident/Problem:** I don't love carrying two devices, and I wish I could have my iPod and phone all in one.

---

1 Jobs, Steven P., Scott Forstall, Greg Christie, Bas Ording, Imran Chaudhri, Stephen O. Lemay, Marcel Van Os, Freddy A. Anzures, and Mike Matas. Telephone Interface for a Portable Communication Device. Apple, Inc., Cupertino, CA (US), assignee. Patent 7860536B2. 28 Dec. 2010. http://1.usa.gov/1GPdPpK

- **Rising Action:** iPhone to the rescue!

- **End:** I now have a way to have my iPod and phone in one device.

This story lacks gravitas or dignity. It has a structure, but when you visually assess the story as a diagram, you can see that the structure isn't very tall. Both the diagram and the concept for what could have been the first iPhone lack two key plot points that give stories their height: crisis and climax. The story is so flat that Steve Jobs mocked the prototype by jokingly announcing a product that looked very similar to the prototype before unveiling the actual iPhone during the 2007 keynote.

Our brains don't like flat. Our brains need structure in order to get excited about things before, during, and after experiencing something, whether the experience involves thinking about or actually using the thing.

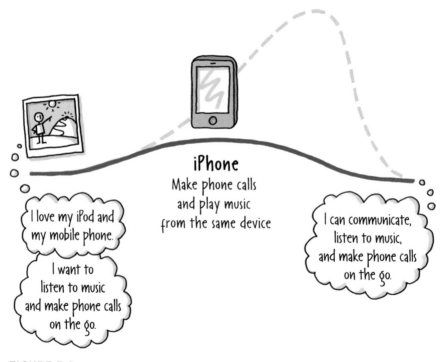

FIGURE 3.6

The anticlimactic concept story for what could have been the first iPhone.

Contrasting what could have been with what actually became the first iPhone helps you see how concept stories aren't just ideas, but rather stories embodied in the things you build.

A simple solution, an iPhone that made phone calls, was flat.

A climactic solution, an iPhone that did everything, worked like "magic."

One was a 2-in-1 device, the other 3-in-1. One helped you play music and make phone calls, the other was a communication powerhouse.

Supporting the Story

"It's easy to come up with something pointless. But in any good storytelling, every scene has a function and a purpose and a point. If it doesn't, you cut it out."

—Vince Gilligan,
Creator: *Breaking Bad*

The conceptual story of the first iPhone bolstered the entire keynote presentation that Steve Jobs delivered. But the story wasn't just a marketing pitch or shiny packaging—it drove everything from *what* the product requirements were to *how* the iPhone worked.

And more importantly: concept stories don't just help you figure out how to *talk* about a product, but how to *build* the product.

There was a good reason why the iPhone launched with fancy animation but no cut-and-paste functionality. Its concept story necessitated an advanced presentation in order to embody the device's value proposition, like communication-enabling apps and a touchscreen. And in order for the touchscreen to be user-friendly (which Palm Treo touchscreens were not), Apple's strategy was to employ animation and cutting-edge technology that enabled people to tap, rather than to interact with their device using a stylus.

Animation supported the story. Cut-and-paste did not.

All of the marketing in the world cannot grow a business or a product line that doesn't deliver on the story that using the device promises.

> If you ever used a Palm Pilot or Treo phone, you'll recall how difficult it was to use a touchscreen without a stylus. It was hard to tap the right touch target on a screen, and there was a delay in the system giving you feedback that you just pressed something. It was easy to get lost from screen to screen and equally difficult to navigate around the system. Palm Pilots and touchscreen smartphones were a novel idea, but they didn't have a huge market, nor were they in high demand. They simply weren't easy to use, and that wasn't a very good story.

Mapping a Concept Story

There are a few different times you might want to map out a concept story for your product. For example, let's say you are working on a brand new product or feature. You could map out a concept story in order to brainstorm or define your product from scratch. Or you could map the story out in order to assess whether or not an existing idea was any good or could possibly engage your target audience.

Let's say you're not working on an entirely new product, but are instead working on a marketing strategy, user flow, app, or website for an existing product (more in Chapters 4 and 5). In that case, before you got deep in the weeds, you would first map out a concept story so that you and your team fully understood what your product was, as well as its core value proposition. Doing so would help you ensure that those elements were incorporated and communicated in each and every story you mapped out thereafter.

In order to map out your concept story, you must answer these questions by plotting them onto your narrative arc.

- **Exposition:** The current state of things:
  - Who is your target customer?
  - What's good in her world as it relates to your product or service?
  - What is her big goal as it relates to your product or service?
- **Inciting Incident/Problem:**
  - What is her problem or pain point?

- **Rising Action:** The name of the product
  - What is the name of your product?
  - What type of product is it?
- **Crisis:** The competition
  - What does the competition look like?
  - What mental hurdles might keep her from adopting your solution?
- **Climax/Resolution:** The value
  - What will help her resolve her problem *and* overcome a crisis moment or resistance?
  - What's your product's primary value proposition or differentiator?
- **Falling Action:** The takeaway
  - What do you want people to think, feel, or envision after learning about your product?
- **End:**
  - What happens when the user meets her goal?
  - This is where the business meets a high-level goal or fulfills its mission, too. What's the business goal? How will you know you're on the right path?

As you can see with the Slack concept story, much like with the first iPhone, there is always a simple solution to every story, which can look like a straight line. People want to communicate and collaborate? So give them a better communication and collaboration tool. However, while you might want to design solutions that are simply *better* than whatever else is out there, "better" isn't always enough. Having a product be "simple" isn't always the most compelling or motivating story.

With Slack, you see something that maps out well within a structurally sound story. The company could have built an online messaging platform instead that was "easier to use" than email, for example. Often, clients and stakeholders on projects ask—what's our requirement? Make it "easy to use!" Or what's the problem? "Oh, our product is difficult to use." But "easy to use" is a pretty boring story when difficulty isn't really the problem.

Here's how you would map out the concept story for the online collaboration tool, Slack (see Figure 3.7). *The answers to the questions are in italics.*

- **Exposition:** The current state of things

  - Who is your target customer? *Busy professionals*

  - What's good in their world as it relates to your product or service? *Communication and collaboration at work is easier and happening more than ever before. They've got more tools to get their job done—i.e., email, Twitter, Facebook, SMS, instant messaging, video chat, online project management software with built-in messaging—a plethora of ways to get in touch, stay in touch, communicate, share, and collaborate with their team.*

  - What is their general goal as it relates to your product or service? *To communicate and collaborate with a group of people.*

- **Inciting Incident/Problem:**

  - What is their problem or pain point? *Communication and collaboration is a pain. While it's easier and more accessible than ever, it's still difficult to keep track of everything or keep it all in one place. This means that what should be easier ends up being harder, wasting their time and money.*

- **Rising Action:** The name of the product

  - What is the name of your product? *Slack.*

  - What type of a product is it? *An online collaboration tool.*

- **Crisis:** The competition

  - What does the competition look like? *Free online services like Gmail or IM. Existing services that people might already pay for like Basecamp. There's a very long list here.*

  - What mental hurdles might keep them from adopting your solution? *What, sign up for another online service? Ack. No.*

- **Climax/Resolution:** The value

  - What will help them solve their problem and overcome a crisis moment or resistance? What's your product's primary value proposition or differentiator? *Unlike the competition, Slack is a one-stop solution. Have your communication in one place. And the best part: they can access it all anywhere, anytime, from their desktop, web browser, phone, or tablet.*

- **Falling Action:** The takeaway
  - What do you want people to think, feel, or envision after learning about your product? *They imagine themselves using Slack with their team-mates…and maybe never using email again. That would be cool.*

- **End:**
  - This is where the users meet their goal. *To communicate and collaborate with a group of people.*
  - Oh, and it's where the business meets its high-level goal, too. What's the business goal? *Broadly, to help people better communicate and collaborate with a group.*

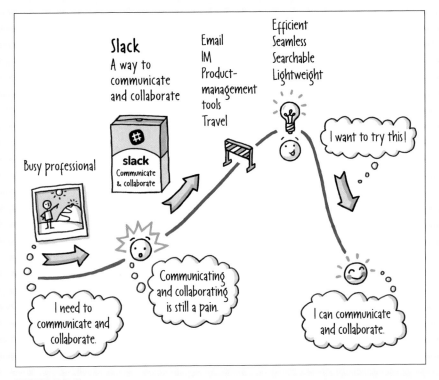

FIGURE 3.7

The concept story for Slack. This is both the story that people hear when they hear about Slack through newspaper articles or word of mouth, as well as the story they remember once they start using the product and experience its value.

In fact, plenty of products aren't *easy to use*, but they have such a solid concept story behind them that customers *love* them. When you build a concept story using this framework, it requires you to meticulously assess and identify the root of a user or potential customer's problem so that you can effectively design a solution that not only gets people excited, but also maps out how their brains see the world.

Finding the Concept Story at FitCounter

NOTE FICTION THY NAME IS FITCOUNTER

"The story you are about to hear is true. Only the names have been changed to protect the innocent."[2]

Although FitCounter is based on my own experiences with real companies and clients, it is solely a product of my imagination. It is *not* a real company, nor a real product. However, as a case study throughout the book, I will write it as if FitCounter were a real company, using details and facts that might make it seem real.

When I first started working with FitCounter,[3] a health and fitness start-up, it had a really great idea and a noble, if not vague, mission: to help people get up-to-the-minute information and news about fitness, sports, and training. Initially, it started out as a Web and mobile-based fitness and workout tracker to help customers track their runs and workouts, but unfortunately it didn't have much success in an already saturated market. FitCounter did, however, produce these great, little, timely exercise, fitness, and health-related videos that people loved. Anytime there was a new trend in fitness, they were on it.

The year before I started working with the company, it decided to realign its tracking tool and revenue model around a new approach: FitCounter would become a content provider, and hopefully, people would sign up and pay for access to its content. Some people signed up, but very few people paid. Things were not looking good.

---

2 Opening trademark statement from *Dragnet,* a radio, TV, and motion picture series.

3 Note: This is a fictionalized version of a business.

The problem was that while the free videos got lots of traffic and views, very few website visitors signed up to become members, and even fewer of the members paid for premium access. Although many video content providers, such as YouTube or CNN, use advertising to build revenue from free content, advertising was not an option in this case. The board and investors did not want the business to get into advertising. Software as a service—that's what we were asked to build, not an ad-driven content platform.

If you're at all confused about this business model, imagine how confused website and app store visitors were. They didn't understand what the product was, why they should sign up, and what paying for a service would get them. The business was pursuing a freemium *software as a service* revenue model, where you try a product out for free at first, love it so much that you use it a lot, and then either pay to be able to use it more frequently or to unlock premium features or services.

But while the business saw a product that people could use, visitors just saw content. And visitors expected this content to be free, like the content they could find on YouTube. Why pay for access when there was probably a decent alternative out there that was free? We had to figure out how to engage users so that they would not only use the product, but also see the value in it and eventually upgrade. If we couldn't figure this out, the business would fail.

What we needed, we eventually learned, was a story—to drive the business, the team, and our potential customers. Actually, we needed a *concept story*.

The story of FitCounter is a perfect example of making the story fit the users. While FitCounter had a hard time acquiring *new* users, it did have a core group of devoted, paying customers—we called them "superfans." They *loved* the product. But, in all honesty, we didn't understand why they loved the product. These superfans logged in to the Web and mobile apps several times per week and spent lots of time using them. We hoped we could—at the risk of sounding completely megalomaniacal and creepy—engineer more superfans like them.

But in order to do so, we needed to fully understand who they were, what their pain points were, what they did with the product, how it solved their problems, and why they loved it. We knew that talking to customers was the first step toward solving this puzzle.

However, what we found when talking to these customers surprised us.

After listening to our customers, testing hypotheses, and drafting and redrafting stories, we eventually realized that FitCounter's concept story looked something like this (see Figure 3.8).

- **Exposition:** The main characters are active, tech-savvy self-starters. Their general goals are to get or stay fit, and they are visual learners who like to use training plans to do so. They love video for fitness training because they can *see* how to do something and follow along as they try out different exercises.

- **Inciting Incident/Problem:** It's hard to find good video training plans because many of them are cookie-cutter applications and not applicable to what customers need to learn or their level. Sometimes, customers come up with their own training plans or work with a trainer to write one, but those plans aren't visual. They often find how-to or training videos on YouTube and make training playlists that way, but the quality is low and the effort is high.

- **Rising Action:** FitCounter is an online training platform that people can use to find good-looking, high-quality fitness videos and playlists, package this content into their own training plans, and share them with others.

- **Crisis:** Is it worth it to go through the effort to use this product to make the training plans? Should they go back to producing their plans manually by scouring YouTube for free videos?

- **Climax/Resolution:** The training plans they create look great—in fact, way better than what they could produce or find on YouTube. And they can personalize them and tailor them for their needs/level. The training plans are made of bite-sized videos that customers can easily fit into a regimen, and they are also easy to share and can be accessed anytime on the go.

- **Falling Action:** Customers see the value in using FitCounter and how it fits into their lives.

- **End:** They can see themselves using the product to get or stay fit and want to try it out.

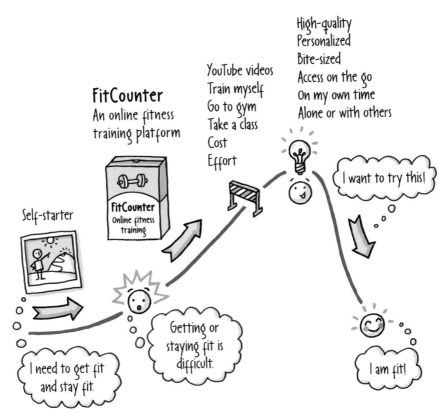

FIGURE 3.8
The concept story that we gleaned from talking to customers that illustrated how FitCounter fit into people's lives.

What we had here was a solid story for a concept for a product—one that we tested by talking to customers, running ads, and measuring with customer feedback, clicks, and eventually acquisition funnels that manifested different facets of this story. At the conceptual level, a product that embodied this particular story was something that people wanted and would pay for.

But this was not the concept behind our current product. That product, again, was video news about health and fitness. It was basically a way to watch exercise videos so that you could get fit. You could watch the videos, save the videos, and you had to pay to access more videos. And sharing was difficult. Finding content? Also difficult because the product was now structured more like a blog with posts, archives, and categories. It wasn't organized around helping people create or share training plans.

Could we, perhaps, build an actual product around this concept? And was this something that people would *actually* use and pay for? For that, we needed an origin story.

CHAPTER 4

Origin Stories

"The character must for some reason feel compelled to act, effecting some change..."

—John Gardner,
The Art of Fiction

In 2011, I got an email inviting me to join the social bookmarking site, Pinterest (see Figure 4.1). The email wasn't pretty. The copy wasn't mind-blowing. Still, this email caught my attention. It called me to action and helped me become a converted and dedicated Pinterest user within a few minutes. This email was an integral part of my *origin story*, or the story of how I started using Pinterest to do things like decorate, cook, and garden.

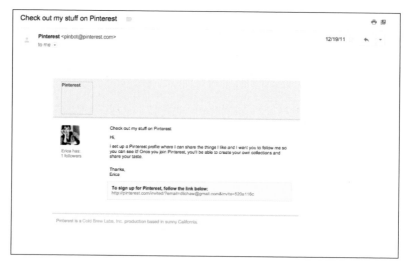

FIGURE 4.1
An invitation to join Pinterest.

While this might not seem earth shattering, this is the origin story that 100 million people around the world have with Pinterest, one of the fastest-growing social networks at the time. How did an otherwise innocuous-seeming email convert so many people? In order to understand how, you need to understand how origin stories function.

What Is an Origin Story?

An origin story is the story of how someone becomes your customer for the first time—it's how that person goes from hearing about your product to actually using it. For you, it occupies the space between how you market a product and design the actual product itself. For your customer, it occupies the space between what they *think* about your product and what they *do* with your product.

As such, an origin story acts as a bridge between your *concept story* and your *usage story*—bridging the gap between the concept of a product or service and the actual usage of it. It's where and when potential customers not only *see* what they can do with a product, but also how they can take an action with it.

Let's look at how the Pinterest email worked to create my personal origin story. First, the subject line of the email was pretty straightforward: "Check out my stuff on Pinterest." Next, in the body of the email, I could see that it was sent on behalf of my partner, Erica. The copy read:

> "Hi,
>
> I set up a Pinterest profile where I can share the things I like and I want you to follow me so you can see it! Once you join Pinterest, you'll be able to create your own collections and share your taste.
>
> Thanks,
>
> Erica"

Finally, there was a link I could click to sign up, which I did without hesitation. But why was I so eager to sign up? A few key words in the email resonated with me and supported what could be my story if I signed up.

Here is the origin story broken down into the story format of a Pinterest user like me, as shown in Figure 4.2:

- **Exposition:** I love to cook and am redesigning my living room. I collect and share things like recipes and home decor inspiration by reading blogs, using my browser bookmarks, emailing them to others, and occasionally tweeting or posting to Facebook.

- **Problem:** Collecting and sharing is a pain. I can never find things after I've saved them. And bookmarking and sharing is typically text-based, whereas home décor and food are visual, so it's hard to find things after I've saved them.

- **Rising Action:** I get an email invitation to join Pinterest. So I joined. What did I have to lose?

- **Crisis:** What? Another thing to sign up for? I already have ways to collect and share things. Maybe I should rethink this?

- **Climax/Resolution:** Someone wants to share something visual with me! It's visual! And social! It sounds simple, but this is a very big deal. All the email has to do is say these two words, and I'm sold: share and see.

- **Falling Action:** I sign up.

- **End:** I collect and share something. I get to experience how awesome this thing is. I'm hooked!

FIGURE 4.2

The origin story of how someone becomes a Pinterest user.

> Life is a story. And in this story, you are the hero. If your custom-
> ers are the heroes in their stories using your product, then just
> like comic book superheroes, they get an origin story—that's the
> story of how they became a hero for the *very first* time.

Origin stories like the one that included the Pinterest email can be
that simple: a few lines of copy in an email that help the business
grow astronomically. Apparently, I'm not the only one who wants to
collect and share visual things. Are you really going to go out and
build the next Pinterest? Probably not. Can you really credit Pinterest's email strategy and design for its explosive growth and new user
acquisition at the time? Of course not. Many factors play into precipi-
tous and even viral growth like this, such as word of mouth, press
coverage, or paid advertising.

But what you can do is reverse-engineer an origin story framework
by looking at successful products and how and why they acquired
users. Then you can consciously engineer, build, and test an origin
story for your own product or project.

How Origin Stories Work

Origin stories operate much like concept stories. The biggest difference is that where a concept story helps you define your product and
its value proposition, the origin story helps you figure out how to
communicate your value proposition, get people to take action, and
get your customer to experience that value proposition for the very
first time. You'll even notice that because your origin story relies so
heavily on elements you uncover and outline in your concept story,
you can reuse some of the plot points.

Origin stories generally work like this (see Figure 4.3):

- **Exposition:** The current state of things (same as your concept story)
- **Inciting Incident/Problem:** The problem or emotional trigger
  (same as your concept story)
- **Rising Action:** The acquisition channels
- **Crisis:** The resistance or impediments that the user experiences
- **Climax/Resolution:** Why the user cares
- **Falling Action:** The user takes some kind of action
- **End:** The goal met—the end…for now

FIGURE 4.3
The mechanics of an origin story.

Exposition

During the exposition, you're introduced to the world of the character or user. You find out who the user is and what he wants (his big goal). What does he need to do?

> **NOTE YOU'RE NOT SEEING DOUBLE (OR TRIPLE)**
>
> The material in the exposition is the same in all three stories: origin, concept, and usage.

Inciting Incident/Problem

The inciting incident is the problem or need that your user has. This might be a problem she knows she has, or one that you need to show her that she has. Both of them are valid. And most likely, it's a problem that she usually solves through other means.

Where concept stories describe why someone might want to use a product, origin stories illustrate how they find it and then why and how they use it for the very first time. *A concept story is big picture, while an origin story starts to get tactical.* Additionally, an origin story provides a complete story for your users with a beginning, middle, and end.

Rising Action

Your rising action is represented by your acquisition or brand awareness channels—for example, how your customer hears about or finds your product. The rising action occurs when your story starts transforming from conceptual to something more like a journey with actual events. Here, your customer might find your product in a variety of ways:

- Hears about your product or service from a friend through word of mouth

- Does a Google search

- Sees an advertisement on TV

- Reads a tweet, Facebook, or other social media post

- Receives an email

Crisis

At the crisis point, the conflict and tension start to build. You might have uncovered potential crises when you developed your concept story. If so, consider and include them here.

Remember, conflict and tension are good; they make stories more enjoyable and satisfying for the person experiencing the story, whether they are watching a movie or even becoming a customer. If there is no tension, then your solution is probably not very exciting.

In origin stories, much like concept stories, these crisis moments will most likely happen in people's heads as they think about why they shouldn't take some kind of action. Or the crisis moment might be something that you uncover and want to avoid, like confusing copy on your website or in an email. Think about your customer's journey. What might get in the customer's way or what hurdles might she face? These hurdles can include the following:

- Competitor products or brands

- Other solutions (like doing things themselves or analog solutions)

- General resistance to trying something new or taking action

- Fears surrounding safety or security

Climax/Resolution

In a concept story, the climax occurs when your customer *understands* the value of a product; in an origin story, it happens when the user *sees* the value of a product. (As you'll see in the next chapter, the climax of a usage story is where the user *experiences* the value of a product.) The user doesn't solve her problem yet or meet her goals, but she sees that she can use your product to do so.

Because this is now an actual step in her journey, you need to consider carefully where this happens. Where do you want her to land after she first hears about you or heads out on her journey? This is the point when all of your story engineering manifests into some kind of interface, whether it be digital, print, or even a customer service script that a salesperson will communicate over the phone. If digital, you've got a few moments at best to grab someone's attention and make him care. These can take place in a variety of ways:

- Home pages

- Landing pages

- App Store pages

- Brick and mortar stores

- Calling a phone number

- And a plethora of other key touchpoints

Falling Action

What then? She just saw that she can solve her problem by using your product. What do you want her to do now? You'll want to consider one primary action that you ideally want people to take—a *happy path*, if you will—but also consider and plan for multiple actions and potentially branching paths. Broadly, you've talked to your customers, looked at your analytics and funnel traffic, and parsed your stories from all of the data, so what primary action do you want this person to take once she is sold on using your product?

Is there a secondary action she might take? You might want her to sign up for something, but what if she just wants to get more information instead? No matter what the action is, this is the point at which you not only want her to take an action, but you also have to get her to experience some kind of value. Nothing says "not coming back" like a sign-up form that takes you to a "thank you" page.

Think like a storyteller. How can this episode come to a close in a way that satisfies both your customer and the business? You don't need to go into too much detail for the falling action. What you might find is that the falling action is probably its own story that warrants closer inspection (more on that in the next chapter).

Here are some common falling actions, each of which merits its own story:

- Try out a demo
- Sign up for an account
- Learn more
- Call a phone number

End

This is where she meets her goals. It's also where the business meets its key goals. The key to developing successful origin stories is that you call your main character to action and determine how to measure that action once they acquire their goal and in every step that leads to the end.

Mapping an Origin Story

After you've drafted a concept story, you're ready to start working on an origin story. Draft an origin story in any project where you are trying to figure out how to acquire or convert customers. Marketing strategies, landing pages, even something as specific as naming buttons and calls to actions all benefit from a clear and structurally sound origin story.

Map an origin story for any and all of your markets, personas, or customer behavior types that you are targeting, as well as the key touchpoints and acquisition funnels you need to explore. For example, the story of how a customer type finds your home page, landing page, or app through a Google search? Map out that story. The story of what happens after someone sees your TV advertisement or Facebook ad and then decides to buy or try your product out? That also gets a story. That home page that isn't converting enough new users? Map out that story, too. Chances are, there is a cliffhanger in there that you can identify and troubleshoot.

When mapping out an origin story, you want to ask yourself these questions:

- **Exposition:** The current state of things (same as your concept story)
  - Who is your target customer?
  - What's good in his world as it relates to your product or service?
  - What is his goal as it relates to your product or service?
- **Inciting Incident:** The problem or emotional trigger (same as your concept story)
  - What is his problem or pain point?
- **Rising Action:** The acquisition channels
  - How might people hear about your product? How does this path toward discovery relate to their pain point? Acquisition channels and paths can include things like:
    - Hears about your product or service from a friend
    - Does a Google search
    - Sees an advertisement on TV
    - Reads a Tweet, Facebook, or other social media post
    - Receives an email
- **Crisis:** The resistance or impediments that the user experiences
  - What might get in their way or what hurdles might they face?

- **Climax/Resolution:** Why the user cares
  - Where do you want users to land after they first hear about you and head out on their journey? These can be places like:
    - Home pages
    - Landing pages
    - App Store pages
    - Brick and mortar store
    - Calling a phone number
  - This is the point when someone lands on a critical screen or step in his discovery process. How will you get him to care about what you have to show him? What parts of your story will you show him to make what he sees resonate? This is the point when all of your story engineering manifests into some kind of interface, whether it be digital, print, or even a customer service script. If digital, you've got a few moments at best to grab someone's attention and make him care.
- **Falling Action:** The user takes some kind of action
  - What action do you want him to take at this point in time?
    - Try out a demo
    - Sign up for an account
    - Learn more
    - Call a phone number
- **End:** The user meets his goal
  - What is the user goal?
  - What is the high-level business goal or mission? It should be something that is measurable (see Chapter 7).
  - How will the user know he has met this goal?

NOTE WHO OWNS THE STORY?

For this reason, origin stories are where marketing, sales, business development, and advertising intersect with product design, and development. Uncovering and enabling a great origin story is not the job of one person or department, but rather an interdisciplinary, cross-functional effort.

Case Study: Slack

Here's how an origin story might map for a product like Slack, the online collaboration tool outlined in the previous chapter (see Figure 4.4)—the answers are in italics.

- **Exposition:** The current state of things (same as your concept story)
  - Who is your target customer? *Busy professionals*
  - What's good in her world as it relates to your product or service? *Communication and collaboration at work is easier and happening more than ever before. They've got more tools to get their job done—for example, email, Twitter, instant messaging.*
  - What is her general goal as it relates to your product or service? *To be in touch with her team.*

FIGURE 4.4

Slack's origin story, reverse-engineered.

- **Inciting Incident:** The problem or emotional trigger (same as your concept story)

  - What is her problem or pain point? *Communication and collaboration is a pain. While it's more accessible than ever, it's difficult to keep track of everything. This means that what should be easier, ends up being harder, wasting your time and money.*

- **Rising Action:** How can she find you or the acquisition channels?

  - What event or events will happen that maps out onto this pain point? *Hears about your product or service from a friend.*

- **Climax:** Why should she care?

  - Where do you want your customer to land after she first hears about you and heads out on her journey? *Home page*

  - How will you get her to care about what you have to show her? What parts of your story will you show her to make what she sees resonate? *On their home page, Slack literally spells it all out for you (see Figure 4.5): "Slack is a platform for team communication: everything in one place, instantly searchable, available wherever you go." In case you don't read (what? never!), no worries, you can see photos of people like you using the product.*

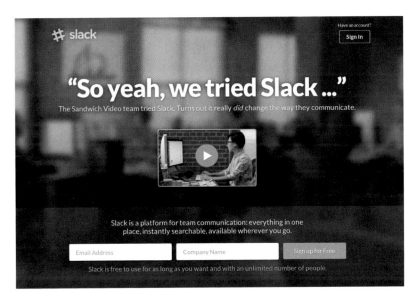

FIGURE 4.5
Slack's home page spells out its value proposition, which is the high point of this origin story.

- **Falling Action:** The user takes some kind of action

  - What action do you want her to take at this point in time? *You want her to sign up.*

- **End:**

  - This is where she meets her goal: *to communicate and collaborate with a group of people.*

  - How will you measure success? *The increased percentage of new sign-ups? The total amount of registered users? Both are valid, but it's important that you choose one as your marker for success.*

If you haven't signed up for Slack or joined an existing Slack group, I highly recommend trying it out. Much like Pinterest, it is a great example of story architecture that doesn't end with a simple sign-up flow and a bored user. Falling action, in the case of Slack, works as its own little story, because a new user is introduced to the concept, prompted to try things out and interact with the service, and sees the value and outcome of what she signed up for. Much like Pinterest, she doesn't just get the promise of communicating, but she actually does it. As with a usage story that you'll learn about in Chapter 5, "Usage Stories," she not only *sees* the value, but also *experiences* it. Just like a good story.

Case Study: FitCounter's Origin Story

At FitCounter, the concept story was something we stumbled upon during research. We initially thought the company was in the business of producing short, up-to-the-minute exercise and fitness videos, essentially maintaining a blog and directory of content. But the business couldn't figure out how to make money off this model without selling advertising, which was not an option. Most people were not signing up for new accounts, nor were they paying for premium content. But what we did know was that a core group of devoted (and paying) customers had a universal story: with FitCounter, they could get fit and stay fit.

Refocusing Our Vision

Once we identified this concept story, it helped us refocus our strategic vision and fully understand what the product was and could be for customers and potential customers. This simple story helped us develop, adapt, and reconfigure our product roadmap to align with it. As a strategic vision, the story was most valuable because it provided us with a foundation on top of which we could build the product anew. The product the way it was built before was mostly unsalvageable. We had to, shudder, redesign it.

Typically, I stay away from big product, website, or software redesigns because they are costly, take too much time, and are too risky. For example, what if you relaunch, and you were wrong about something? Next thing you know, data shows that you're not meeting your goals—or worse yet, performing worse than before—and the product needs to be fixed all over again.

In this case, however, we decided to redesign the product because it didn't feel as risky as it could have been. After all, we had a story, and it was a good story that a core set of passionate customers had uncovered for us. Plus, it was a story that we had tested and validated with a new set of potential customers to make sure that we were on the right track.

Our concept story helped us find a strategic vision and direction, as well as feeling more confident about our overall direction and product market fit. But even though redesigning the product felt less risky, it was still risky. It would take months, all of our budget, and we could fail, which would mean going out of business.

Testing Our New Vision

To mitigate this risk, we decided to build, prototype, and test our new story in a tangible way so that we could validate it on a larger scale than with small in-person tests. If our superfans found value in what they could do with this product, we hoped that this story would resonate with a larger group of people.

After conducting a series of small tests, like running Facebook ads to see what people clicked on and radio ads to see what drove new customers en masse, we felt confident enough to start our redesign with the smallest body of work that would give us the highest impact: redesigning the product's home page. We wanted to see if we could utilize our concept story to increase our conversion rate and get more people to sign up for the service. We didn't care about them paying to upgrade to premium at this point; we just wanted to know if they would sign up. Could this be a viable ending to a potential customer's origin story?

If our home page experiment worked, we would see if we could engineer our story into everything from branding and identity to marketing and messaging, to content strategy and product strategy, as well as the actual product itself. While doing all of that was still risky, it would be less risky knowing that the story was validated on our home page. And if we couldn't validate this story on our home page, we would figure out Plan B. But we didn't think too hard about that because we were confident in what we were about to do.

The Plan

With our concept story in hand, we set out to map our origin story and build the front door to an online training platform that communicated everything our superfans told us they needed. This platform would help people train and get fit. This product was no longer a way to watch videos about the newest, shiniest running trend. Instead, it was a way to learn how to train yourself or others for something like running a marathon. Before we could build a product that did that, we needed to communicate to a new set of visitors what they *could* do with this product if they signed up to use it. For that, we drafted an origin story.

What we needed to determine at this stage of product development was the following information:

- **How would potential users (we called them *visitors* until they signed up) find us?** In other words, what were all the possible channels and touchpoints where they could first access us? For example, did they find us through word of mouth and direct traffic to our home page or an App Store landing page? Did they look for something specific in a search engine and first encounter the product by landing on a lower-tier video page? All of the above? What else? We outlined all of the journeys and scenarios we could think of.

- **What value should we communicate at each of these points of entry?** Whatever we previously communicated either didn't work to get people using the product or if it did, it just confused them afterward. We decided to tell people directly why this thing was awesome. Plain English. Or maybe other languages? Evaluating channels would help us figure that part out.

- **What affordances should we show?** For example, what should people see that they could do with the product? (And we didn't just mean buttons, although buttons might be a solution for calls to action.)

- **How should this first-time encounter story end?** People knowing what they could do with a product wouldn't save the business. We needed to figure out what the first action would be that we needed them to take. Did we want them to sign up, use the product, convert to premium, or all of the above? Each ending had a different strategy associated with it and we needed to consider them all.

The Cliffhanger

In order to answer these questions and determine our origin story (or stories), we started by looking at some quantitative data. We involved our SEO team and analyzed search engine queries and traffic so that we could see what kinds of keywords people searched for when they first found us. We also looked at our Web analytics. Did people find us through word of mouth and go directly to our home page? Or did they find us through Google or social media sharing? We needed to identify the stories our data was telling us. Were new users having experiences with our product that meshed with the stories we heard from our superfans?

Luckily, the data substantiated what our superfans told us: most of our visitors found us by searching Google. Instead of wanting to find out things like who won the NYC marathon (like we initially thought), they were searching for things like "how to train for a marathon." We also saw this trend in our internal Web and mobile analytics. By far, the most highly consumed content was training related, not news or special interest. People were using the product to train, but the product never told them they could do so. *We needed to change that.*

The story that our data showed us was that while our superfans had a complete story with our product—one with a beginning, middle, end, inciting incident, climax, etc.—our data illustrated an incomplete story. Here's how the origin story played out for someone finding the home page for the first time (see Figure 4.6):

- **Exposition:** People want to train themselves or someone else, and they prefer to replace or just supplement the in-person interaction with online training.

- **Inciting Incident/Problem:** Making high-quality training videos or finding them on YouTube and then compiling them into a series is a pain. There must be a better way.

- **Rising Action:** They find the FitTracker on Google by searching for something like "how to train for a 5k" or "fitness training."

- **Crisis:** They go to the home page, and don't see what they're looking for, so they bounce. Instead of seeing something that tells visitors that they can use this product for fitness training, they see a website that asks you to sign up to get up-to-the-minute news on sports, fitness, training, and exercise.

As you can see, this story is a cliffhanger, because it ends at the moment of crisis near the top of the arc.

FIGURE 4.6

The FitCounter cliffhanger: people signed up, but rarely used the product.

This cliffhanger appeared in other origin stories, too. Like the one with the email invitation as the rising action. Or the one that took place in the Apple App Store. Or the one that happened when someone would come across a single video page, watch a video, and leave. All cliffhangers. We needed to complete this origin story, not just for the business, but also for our potential customers. Without a complete story, the goals weren't being met.

The Story

Based on our customer interviews and quantitative data analysis, we hypothesized that potential customers wanted online fitness training. And we knew that we could offer that if we redesigned the product. But we just needed to make sure that we were on the right track, starting with how a potential customer would first convert to being a registered user. Just like a superhero.

What we needed to design for was an origin story that looked more like what you see in Figure 4.7 and let people do these things:

- **Climax/Resolution:** I see that I can use FitCounter to get and stay fit. In particular, with FitCounter, I can get personalized training plans that utilize bite-sized, high-quality videos. I can train on my own time, anywhere, and do so alone or with others.

- **Falling Action:** Sign up to try it out.

- **End:** Get fit (and help others get fit).

Unlike before, our story was now complete. The story contains all of the values and high points that we outlined in our concept story based on what our superfans told us. Our hope with this was that if people experienced a story like this, the first-time encounter with the product wouldn't be a cliffhanger for them or for the business. Once we had a clear idea of what our origin story was, we set out to redesign the home page so that visitors not only knew what to think about our product, but also what they could do with it. As you can imagine, this is when our marketing, product, and design teams all became best friends as we aligned toward a single vision and a single story.

FIGURE 4.7
An ideal (and climactic, we hoped) origin story for FitCounter.

The Solution

Outlining an origin story is only the first step toward building a path to discovery and conversion (or whatever your goal might be on a given project). We needed the home page to embody this origin story. And then we needed visitors to experience this story. Much like any design project, we set out to build a new home page and measure its success. We wanted to see if potential customers' words (we would test designs out in-person with potential users) and actions (in this case, clicks and conversion rates) mapped out to an origin story that we heard and believed could happen for more people than just our superfans.

As you can see in Figure 4.8, the content for the former home page tells you that you could "sign up" and "stay up-to-date on all things training and fitness." But, as we now knew, this is something that very few people want or need to do. What you can see instead in Figure 4.9 is a schematic for a home page that embodies a different story for a visitor. It literally spells out what they can do with it: "Get fit, stay fit, and start training."

FIGURE 4.8

A rough outline of the former home page.

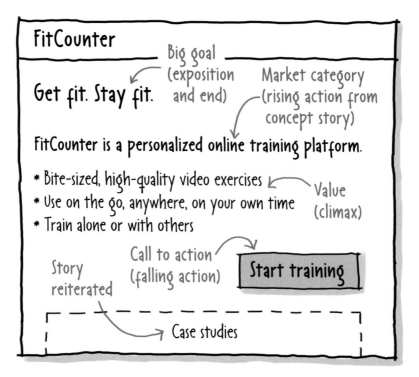

FIGURE 4.9
An ideal (and climactic, we hoped) origin story for FitCounter
(the new rendition).

In addition to clearly communicating what visitors could do with the product, as well as its value propositions, it echoed this story with testimonials from our...you guessed it: superfans. Our marketing and content teams wrote words that communicated a story, and we collectively chose our users' words so that they could tell their own stories of how they got fit using FitCounter. As a visitor, you could now see what your story might be: get fit, stay fit.

The Results

This home page is not perfect, and it is something that the business still tests and iterates today, but as we found out, it was a pretty great start. Good enough to get 40% more people to click on the sign-up button than before, which is something we had never been able to do previously. And we did that by building around an origin story that we first heard from our customers and then repeated for a broader set of people. We built this page story-first, as opposed to being built on best guesses and intuition. Finally, our site resonated with people.

In this case, qualitative and quantitative data were essential in diagnosing, understanding, and hypothesizing FitCounter's origin story. We didn't just listen to our superfans and build what they told us. Rather, we saw their stories echoed in our analytics as we looked at usage data, traffic patterns, and funnels. We heard a story from our superfans, posited it as an *educated* guess (hypothesis informed by real life data), then prototyped it, and tested it to see how it performed.

We also used the origin story to determine requirements that supported the story (more in Chapter 7). For example, we outlined that the home page needed to do these things:

- Include photos of people doing things that matched our primary persona's behavior, using materials we gathered during our initial research when we found our concept story. These were regular people who wanted to get fit or stay fit, not professional athletes with already toned and chiseled bodies. It was important that the imagery be identifiable but obtainable.

- Communicate persona goals as something attainable: they could get fit or help others train and get fit.

- Clearly articulate what our target personas could do with the product. The page's primary call to action should reflect what they *could do*. Visitors rarely sign up for a product or service as a goal in life. The call to action like "start training" maps to the potential users' origin story, whereas "sign up" does not.

- Search engine optimization: now that we had a better idea of what people could use our product for, we could optimize for queries like "fitness training" or "how to train for a marathon."

We also used the framework of an origin story to plan for, design, build, and test ideas for many different contexts, user types, channels, and paths with similar results. We drafted origin stories and requirements for journeys involving the App Store, deep-linked back doors (directly to video content), social media sharing, paid advertising, and email invites. We also drafted and tested origin stories for different types of personas to make sure we accounted for different goals. We even used our master origin story as the foundation of a radio ad that performed extremely well. In developing a product with a solid story structure, not only did we make new customers happy, but we also made our ad agency happy.

We went not just story-first, but story-crazy. In the best possible way.

And for good reason: it worked. Building story-first helped us figure out what we were as a concept and how we fit into our users' lives upon first contact. Not only could we better understand what the product was, but we also knew what the story of someone discovering and seeing potential in *using* a product could be.

But as we knew would happen, without a fully fleshed out product behind the front door of the home page (or any of the other back doors), our origin story was incomplete. In other words, we finally validated that people would sign up to use a training platform. But once they signed up and tried the product, they didn't complete their journey.

Instead of getting fit or giving the product a test run and seeing that they could potentially get fit if they kept on using it, they struggled to use the product. Behind the front door, the product was still a platform for getting up-to-the-minute news on fitness and training. We solved our first crisis and resulting cliffhanger of people not signing up, but we now had a new one to solve: how to build an online training platform that let people get fit or train others and help them get fit. For that, we moved onto *usage stories*.

CHAPTER 5

Usage Stories

"The same core features appear in the rules of narratives and in the memories of colonoscopies, vacations, and films."

—Daniel Kahneman,
Thinking Fast And Slow

In the 1970s, researchers conducted a series of experiments on how humans experience pain. Their subjects were patients who underwent colonoscopies. While the technology and overall procedure is now less painful than it used to be, back then it was not just unpleasant, but extremely painful. What the researchers wanted to know was if the *duration* of a procedure affected the overall experience. In other words, if a painful procedure was twice as long, did a patient consider it to be twice as painful? Or if it was half as long, was it half as painful? The researchers learned that while duration plays a slight role, other factors play a bigger role in shaping experience: peaks and ends.

The researchers used self-reporting mechanisms to record how patients felt both during and after the procedure. Test subjects were asked to rate their pain level on a scale from 1 to 10 on regular intervals, 1 being the least and 10 being the most painful. After the procedure, they were then asked how it was overall and how likely they were to choose to have the procedure again in the future. Researchers assumed that the more painful and longer the procedure, the less likely a subject would want to repeat it.

It turns out, however, that if two patients rated their pain level as consistently high throughout the procedure, the patient with the shorter procedure was no more or less likely to rate it differently than the other patient. Both rated it as awful and were not likely to want to repeat it. If a patient experienced extreme pain for three quarters of the way through the experience and then felt that pain gradually ease until the procedure ended without pain, the results were drastically different. That particular patient was more likely to rate the overall pain level as being lower than the other patients. And these patients were also more likely to say that they would have this procedure in the future.

What researchers extrapolated from this and other studies like it is that humans remember, not duration, but rather the peak of an experience and whatever happened closest to the end. This phenomenon is called the *peak-end rule* (see Figure 5.1). A peak can be painful, as in the case of a colonoscopy, or it can be enjoyable as with a vacation, a film, or—gasp—the experience you have using a website, app, product, or service. Everything—even the experience you have when you use billpay through your bank's website or app—is a story. It's up to you as someone who designs or builds things to determine if that story is going to be a good one or not.

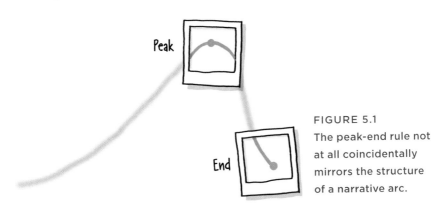

FIGURE 5.1
The peak-end rule not at all coincidentally mirrors the structure of a narrative arc.

As Nobel prize-winning psychologist Daniel Kahneman, who worked on this study, points out, there is a reason why our brains have evolved to give shape to how we comprehend and communicate experiences. There is no such thing as an *actual* experience. All you have are moments in time. A moment ago? It's already gone. The moment you took to read this sentence is already gone by the time you reach the end of the sentence. Time is fleeting. All you have is the memory of events that happened in time and the ability to stitch together and parse meaning from those data-points into a coherent narrative.

This real-time processing is what sets apart humans from other animals. The foundation of this cognitive function is story. And story has structure. Story is how you make sense of the world around you—before, during, and after an experience. When you consume a story, whether it involves listening to a story or parsing life as a story in real time, your brain is activated. If what you just experienced maps out to a narrative structure, with a beginning, hook, middle, peak, and end, it maps out to how your brain is pre-programmed to understand the experience. When you experience something like a story, it affects comprehension, utility, perception of usability, memory, and choice. In other words, you're more likely to understand something, see it as useful, find it easier to comprehend or use, remember what you just experienced, and want to repeat the experience again. Even painful medical procedures. Or sign-up flows.

What Is a Usage Story?

Usage stories are exactly what they sound like: the story of someone using your product or service—step by step. It's the actual steps that make up the story for your user, plot point by plot point. The steps in a usage story can involve screens, if you're working on a screen-based website, app, or software. Or the steps can be things that happen outside of the screen if you're working on something that is entirely non-screen-based, such as an experience strategy for a university welcome center.

More than likely, anything you build a usage story for will be a combination of screen and non-screen-based steps. For example, that university welcome center might have screen-based kiosks that help visitors find what they're looking for, as well as signage and other affordances—oh, and humans who might hang out at a service desk. That app you're building might have a user flow with steps that take place outside of the screen, such as when a customer calls customer service for help. It is essential to consider usage stories within their broader context of who, what, when, where, and why someone is doing something. And, of course, it is essential to consider, assess, and plan for the intended story of use. Plot point by plot point.

How Usage Stories Work

You can employ usage stories to figure out how to structure journeys, long and short. A usage story can take place over a period of seconds, minutes, days, weeks, or years. They help you figure not what a customer should think about your product or how they find your product, but how and why he will use, experience value in using, and continue to use your product in one sitting or over time.

Just as with concept and origin stories, your usage stories can be based on real data or sketched out as a hypothesis. For example, if you are troubleshooting a checkout flow and want to figure out why people add items to their cart but rarely check out, you can use real data. You might get that data from your web analytics, in-person user interviews, usability testing, or all of the above. If you're inventing a product from scratch and figuring out how a key flow works for the very first time, you might base it on data gathered from market or user research or use stories as a way to think outside the box and get creative with how you envision your product working. Or both.

Here is how they operate (see Figure 5.2).

- **Exposition**—current state of things (same as your concept and origin stories)

- **Inciting Incident/Problem**—event, trigger, or call to action (CTA)

- **Rising Action**—a series of steps

- **Crisis**—potential hurdle or hurdles

- **Climax/Resolution**—the high point when value is experienced

- **Falling Action**—then what? Final step in the flow

- **End**—the end…for now

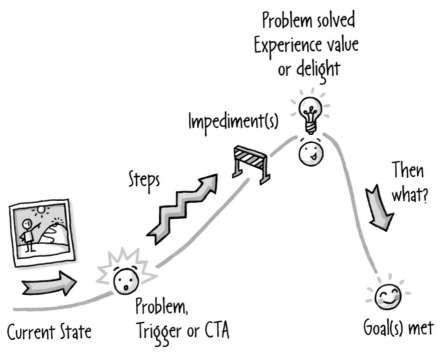

FIGURE 5.2
The model for a usage story.

Exposition

Exposition represents the beginning of the story. There is a main character with a broad goal (which is the same as your concept and origin stories). Where are they at the beginning of the story? Are they using an app, website, in-person service, or all of the above? If you are working on a journey for a business that employs all three as part of their customer service strategy, you might keep your story high level and consider how it works for all three at the same time. For example, the high-level story for someone who is thinking about applying to college is the same whether she is working with a guidance counselor at her high school or doing online research alone. Your character is someone, and she wants something. Does she want to find the right school so she can become a veterinarian someday? Or does she want to find a school that will let her explore career opportunities? Or does she not care about career opportunities and want to study with like-minded people? Does she want to learn more about your school or apply to school? Each exposition is the beginning of a very different story.

Inciting Incident/Problem

The inciting incident is an incentive, trigger, or call to action—something to kick-start this journey. This step should map out onto the problem you outlined in your concept story. If your character is on your home page and her goal is to find a school that will let her explore her interests, how will you kickstart her journey? First, you want to remind yourself what her problem is. Why can't she meet her goal? Because she doesn't know where to start and is overwhelmed by the post-secondary educational landscape, options and opportunities? This is the point where you need to get her to look, listen, and take action. How will you do that? Or if her goal is to apply for the next calendar year, how will you kickstart that journey? As you can start to see, there are many journeys that your user will go on throughout their lifetime of engaging with your product or brand. Each gets its own story as you figure out how to help people meet their goals.

Rising Action

The rising action represents a series of steps or actions the person must take to meet his goal. Each step should build the user's interest and become more interesting or relevant than the last step. This is where the Y-axis of a structurally sound story is especially important. Things don't get good or bad for your user. They get better and better. Just like a good movie, your user should want to continue onto the next step, screen by screen. Action rises and tension rises as your user gets more and more engaged as he tries to meet his goal.

Crisis

The crisis is the impediment that must be overcome for the user to get to the high point of this experience. Impediments can include things like the following:

- Requiring sign-up

- Requiring payment or sensitive billing information

- Mental hurdles, such as boredom, unmatched mental models, or a lack of value

- Poor usability

- Other mysteries—sometimes analytics show a drop-off in funnel completion, and it's necessary to do some research and story-mapping to figure out why people are dropping off

Climax/Resolution

The problem is solved, and the crisis averted. What matters most for a usage story is not just that the problem is solved, but how it is solved. This is where the user experiences value or just feels good about what he is doing. It's the high point of this flow. If just solving a problem is awesome enough, the story will flow well, yet be anti-climactic. Sometimes, however, you need an extra flourish to raise the action level of this plot point to make it more memorable. This "raising of the bar" can be something as simple as an animation for a digital flow, amazing customer service, or a gift for a non-digital service flow.

Falling Action

Falling action occurs when the user finishes the flow. The flow can't just end on a high point—it has to go somewhere and take the user with it. In the case of a sign-up flow, for example, imagine if it ended with a Thank You page. What then? Ask yourself that question every time you build a usage flow and be sure to figure out how it should end.

End

In the end, just like with a concept and origin story, the user's goal is met. The flow is over, and the user should be in a better place than when he started. If you intend for this story to continue at this point, you can consider this stage to be where the goal is met...for now. Just because the hero saves the village doesn't mean that he won't have to undertake a new journey in the next episode. Additionally, in a good story, the main character never just ends up at home after a journey, he should have learned something, found something, or generally grown as a character so that when he arrives back home, he is changed forever and is closer to meeting his big goal or goals.

Case Study: Twitter

One of my favorite examples of a usage story is Twitter's former sign-up flow. While they have since updated this flow, I like using this illustration because it is an excellent example of story structure supporting a flow of screens and interactions. Additionally, this sign-up flow was responsible for not just activating hundreds of millions of new users, but also users who were more valuable and likely to stay engaged with the service over time. While it was not explicitly engineered as such, this on-boarding flow reads (and functions) like a good story.

> **NOTE** TWITTER: ORIGIN OR USAGE STORY?
>
> A few of my workshop attendees always ask why this Twitter case study isn't an example of an origin story. I categorize it as a user story because while an origin story would help you figure out how someone goes from *thinking* about your product to *using it*, this flow simply illustrates how someone uses it. For the very first time. As such, this usage story is the falling action of someone's origin story with Twitter.

The Problem: Low Repeat Engagement

Several years ago, Twitter had a problem: it was starting to grow its user base at a steady clip. But unfortunately, Twitter acquired many new users who tried the service once and then never returned. Twitter's research team talked to users who did return to find out why and what mattered to them.

The answer: people were more likely to log in and engage with the service if they followed others on Twitter who were in their social and professional circles or related to their hobbies and interests. While Twitter's previous sign-up flow was simple, fast, and friction-less—it was only three steps—it didn't do enough to help new users see the value in the service so that they would return.

The Solution: First-Time Use as a Story

Often, there is a rule of thumb that you want to design frictionless experiences so that people get through a flow or process more easily. The easier something is to do or use, the more quickly people will get through it and the more delightful (or less painful) the experience will be, this line of reasoning goes. *Make it easy to use!* is the phrase that your client or stakeholder might outline as a requirement for that flow in the app that you're building anew or redesigning for better conversion. If you think about a usage flow as a story, however, you can see that friction is a good thing. If you think about scientific studies on painful medical procedures, you can also see that shorter isn't necessarily better. What the sign-up flow that Twitter eventually came up with shows is that making something more difficult and longer can be better as long as it reads like a story. Here's how the longer, more complex sign-up flow breaks down as a story:

Exposition

You're visiting the Twitter homepage, which means that you want to know what this Twitter thing is all about (see Figure 5.3). Twitter as a business has the flipside of that goal: it wants to *show* you what Twitter is all about. It also has a more specific goal: to get you to follow relevant accounts so that you are more likely to return in the future.

FIGURE 5.3
The first screen in the flow sets up an exposition: it reminds you why you came here and then incites you to act with a call to action.

Inciting Incident

You see that Twitter is a way to "Start a conversation, explore your interests, and be in the know." Cool, those are all things that you'd like to do. You sign up.

Rising Action

First, you are introduced to the concept of a "tweet," as seen in Figure 5.4. You are on a screen that looks much like what the Twitter app will look like when you are finished. Only there are instructions on the left sidebar and a not-yet-populated area on the right. Someone named the "Twitter Teacher" explains that what you're looking at is a "tweet." You can also see that there are many more tweets awaiting you. You click on the "next" button.

FIGURE 5.4

The first step in the rising action of Twitter's sign-up flow is to learn about a tweet.

Next, you are introduced to the idea of your "timeline" (see Figure 5.5), which you can "build." If you click on a person on the left, you can see one of his tweets show up in the timeline on the right. Click on another, same thing happens. This is how Twitter works—you follow people, and their tweets show up in your timeline. But Twitter doesn't just tell you all of this; you have to actually *do* it a few times before you can go on to the next step.

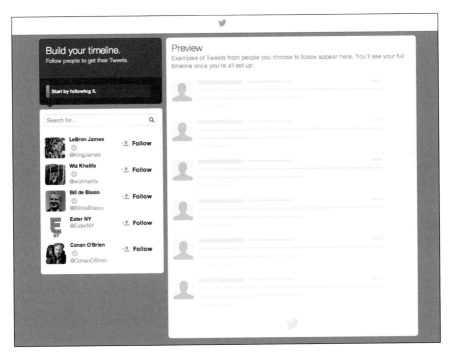

FIGURE 5.5

In the second step of Twitter's sign-up flow, you learn about a timeline.

Now Twitter invites you to "see who's here," as shown in Figure 5.6. While it asks you to "find and follow well-known people," what it is also illustrating is that there are different types of people to follow, depending on your interests. Even though the previous step let you follow celebrities, now you can have a little more control over who those celebrities are—basketball players, for example. The more you follow, the more your now empty timeline fills up again.

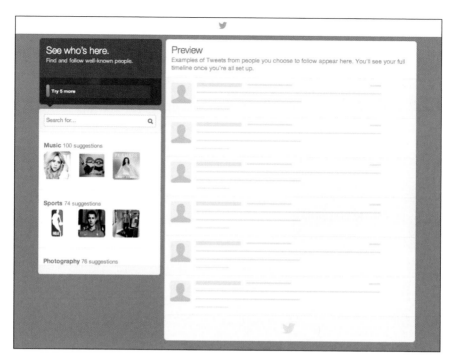

FIGURE 5.6

The third step of Twitter's sign-up flow shows you how to find and follow people.

Crisis

At some point, after spending a few minutes on this sign-up flow (they claim it only takes 60 seconds, but I never completed it that fast), which is eons in Internet time, you might start to get bored. How many more people do you need to follow? How many more interests and hobbies can you think of? You understand what a tweet and timeline are and how the system works. Why follow more people? Not only that, but is reading tweets from Mariah Carey and Brad Pitt all that Twitter is about? You hit the next button.

Climax/Resolution

Now, Twitter invites you to find people you know (see Figure 5.7). My friends are here? Oh, OK, Twitter isn't just about following celebrities. I can also follow my friends and see what they're up to. That's how Twitter helps me be in the know.

FIGURE 5.7

In what could be a crisis moment of Twitter asking you to add contacts, the climax of the story is when you see that your friends are there.

Mini-Crisis

I want to see which of my friends are on Twitter, but is this secure? I don't want Twitter to spam my friends or steal my contact information. As I scan down past the call to action buttons on the left side of the screen, I come to a block of text that allays my fears: I can see exactly how this works. It's safe and secure. Good. I search for and add my contacts from Gmail or wherever.

Falling Action: Add Character

I've learned about tweets, followed celebrities, expressed some of my interests, found my friends, and filled out a timeline. I'm part of this community now. I see everyone's smiling faces and the things they say. I may as well add my avatar and a little bio (see Figure 5.8). I know it should be short because tweets are short. Unlike all of the previous steps in this sign-up flow, I can skip this if I don't want to add my information.

What I don't know is that Twitter *already* got me to do everything it needed me to do—filling out my personal info is nice to have, but not a must-have. Again, what matters is that I understand how "following" works, and I can see the results of following people. At this point, I can add my information or skip it. Falling action like this can be its own story, as with the end of Slack's on-boarding flow (see Chapter 3, "Concept Stories").

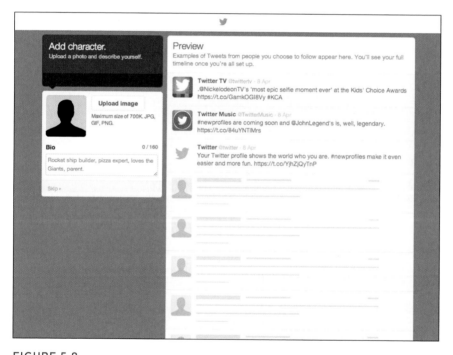

FIGURE 5.8

For the falling action, you can personalize your profile. This part is optional, however, since you've already experienced the value of Twitter: it's social.

End

I'm home (see Figure 5.9). Logged in home. This home page is different than the home page where I started this journey. Before I signed up for an account, the home page *told* me that I could be in the know. After signing up for an account, the home page *shows* me what it's like to be in the know. I see a timeline full of tweets from people and organizations that I'm interested in about topics that I care about.

FIGURE 5.9

At the end of this story, you arrive home (logged in home). But home is better than when you started (logged out home).

What becomes obvious in the story of Twitter's on-boarding flow is that it never once prompted you as a new user to actually try sending a tweet. Instead, at the end of the journey, when you end up home, you can just sit back and consume what's in your timeline without ever sending a tweet. This is a story of following people, consuming content, and being in the know. The story supports not only your user journey, but also a strategic business goal: to get more people to consume content. As Twitter transitioned from a service based on broadcasting to consumption, the story followed along. Traditionally, people are more likely to consume rather than generate content using social media services like Twitter. This flow and resulting story has optimized the story that users are more likely to have and that the business wants them to have.

While the old flow converted new users, the new flow converts quality users—ones who are more likely to engage. You can think of the previous, shorter, simpler flow as the flat or anti-climactic version (similar to the iPhone), or the simple solution for getting people to sign up. The updated flow, on the other hand, is the one that reads more like a story that resonates with the right type of user in the right type of way.

In the end, Twitter is a way to "start a conversation, explore your interests, and be in the know," as millions of people who signed up for the service using this flow found out. Even though you may not have started a conversation yourself, you can see a timeline full of conversations from people whom you are interested in. By the end of this sign-up flow, you're in the know and have had an engaging time getting there.

Mapping the Usage Story

Usage stories can be broad or compact. I recommend starting out with the broadest first and working your way down until you've uncovered all you need. In my experience, each usage storyline you develop should be as simple and straightforward as possible. At the point that you're mapping out many subplot points and getting complex, you should ask yourself if you can break your storyline into smaller stories. Then you should address each separately, starting with the largest first.

Imagine if a TV writer tried to plot out a season-long storyline at the same time as a specific scene from a specific episode. As you can imagine, it's difficult to stay high- and low-level at the same time. Focusing your scope and approaching one storyline at a time will not only improve the quality of your usage stories, but will also help you move faster as you retain focus.

To create a usage story large or small, answer these questions:

- **Exposition:**
    - Who is your target customer?
    - What are her goals as they relate to your product?

- What is the problem or impediment standing in the way of this person meeting her goal?

- **Inciting Incident:**

  - What will kick-start the customer on this specific journey? This will probably be some kind of call to action or event.

- **Rising Action:**

  - What is the first action that the user should take?

  - What's next?

  - And next?

- **Crisis:**

  - What might get in her way of solving her problem and meeting her goal? It could be something tangible, like a paywall, or emotional, like boredom. For each usage flow, you're likely to have a list of possible barriers.

- **Climax/Resolution:**

  - What is the high point of this experience or flow?

  - How will her problem be resolved?

  - What value do you want the user to experience during this flow?

  - What will make all of her conflict, crisis, and work she puts in so far be worthwhile? Value can be functional or more abstract, like brand value.

- **Falling action:**

  - What then? Humans like closure. You don't want to just leave them at a high point and suddenly end the story. Now that her problem is solved, how will you wrap this episode up as quickly as possible so that the user is that much closer to meeting her goal?

- **End:**

  - Where does the user end up—both in terms of character development and logistics? How has she grown? What has she learned? Where is she? What's next? Is this really the end or perhaps the starting point of her next story?

How Big Should Your Story Be?

Stories can be big. And they can be small. They can happen one at a time. They can happen in serial. There is no right or wrong way to scope out the timeline for your stories. Sometimes, you'll scope out a very large story that lasts over a period of years in your customer's journey with your product. Sometimes, you'll focus on a tiny story that lasts only a few seconds. Scope your story so that it answers whatever questions you need to answer.

Maybe your question is "How do we get customers to stay active after a few years so they don't keep dropping off?" In that case, you'll map out an epic journey that lasts a few years. Maybe your question is "How do we get people to come back next time? And the time after that?" In that case, you'll map out something that looks like a serial or a soap opera. Maybe your question is "How do we get people to keep coming back and using our product to do this one core task when there are competitors out there that also do the same thing?" In that case, you'll map out a micro-story. In each case, you have the same plot points and overall structure—only the timeline is different.

Epic Journeys

Epic journeys take place over a long period of time. That timeline can be a day, a week, months, or many years. These journeys are epic because they traverse single sittings or interactions with a product. Maybe you want to map out the lifelong journey from when someone starts using your product to infinity. For example, I worked on a project where we needed to assess the content for a nonprofit educational program—all of its content.[1] That was a lot of content! They wanted to go digital and didn't know what to digitize, what to keep, and what to get rid of.

When we mapped out the user journey over a period of seven years to figure out what content needed to be digital so it could support the story, we saw that there was a lack of content supporting the story around year five, in general. This coincided with a drop-off the nonprofit saw with member engagement, which usually happened around five years after starting to work with the program.

---

1 For full case study co-written by Lis Hubert, see *Storymapping: A MacGyver Approach to Content Strategy, Parts 1-3 UX Matters* http://bit.ly/1k2JckM

Structuring this journey like a story enabled us to quickly assess strengths, weaknesses, and opportunities in their content and program structure. What it also helped us do was to put an "end" point to the member journey so that we could start to envision ways to help members start a new journey with the program after five to seven years. What we found was that after five years, many members dropped off, while others instead became evangelists for the program, disseminating information and training new members. Mapping the journey helped us see that we could plan for not just one epic that lasted five to seven years, but a second story that could last another five to seven years for the user.

Serials and Soap Operas

Sometimes, you will explore your stories in serial. What treating stories in this fashion helps you do is see the relationship between single interactions with a product as a series of stories (see Figure 5.10), rather than just an epic. Doing so will help you figure out how or why to hook people in, get them invested, and stay invested over time as they want to find out what comes next or see what they missed since the last time they tuned in. Just like a soap opera—or *The Wire* or *Breaking Bad,* if that's more your thing—serials are great ways to visualize, plan for, and troubleshoot long-term engagement with a product or service.

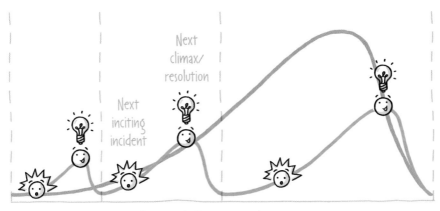

Big goal: Get fit and stay fit

FIGURE 5.10

In a serial narrative, there are mini-stories or episodes that string together over time. Each requires its own beginning, middle, end, and plot points to move the action and user forward.

What visualizing serial stories helps you see is that stories as episodes need to get better for your users over time. If they don't get better, it's much harder to keep users engaged over the long term. Treating stories in this matter also helps you visualize the relationship between different *episodes* or stories.

For example, I once worked for an organization that put on annual events like conferences, workshops, and online seminars. We used serial structure to assess why devoted repeat-attendees attended the first time and chose to return...or not return. The business was selling lots of tickets to its events, and the owners wanted to know if it could do even better. Mapping customer journeys as serial narratives helped us see clearly why people attended (problems and inciting incidents), as well as the value and takeaways that these customers got year-to-year. Mapping serials also enabled us to see the cliffhangers where customers dropped off, as well as gaps and opportunities where the business could add value not just as a whole, but episode by episode.

Micro-Stories: Core Tasks

No flow or task is too small to be treated like a narrative. If you're designing a core task for a system and want to make sure it's memorable and wows your user, thinking in a narrative fashion is a great way to add a layer of complexity and excitement to an interaction— even a simple, seemingly trivial one like adding a calendar event.

The most effective story-driven core task interactions I've seen don't happen during activities like checkout flows. Instead, they are core tasks that are central to a product or service, like sending a message or "liking" something on Facebook. Core tasks are interactions that you want and expect users to perform over and over again.

Consider the iOS Calendar app that comes standard with every iPhone. The story of performing the core task of entering a calendar event works like this:

- **Exposition:** You're having coffee with Jane at 2 p.m. tomorrow. You want to make sure you remember this event.

- **Inciting Incident:** You tap the "Add Event" button to add the event to your calendar.

- **Rising Action:**
  - You type an event name.
  - Add a location.
  - Select a time.

- **Crisis:** Tap…tap…tap…you have to tap so many times to complete this otherwise simple task.

- **Climax/Resolution:** You tap to save the event, and the screen slides down out of view to let you know that the action is complete. It's valuable to know that the system is saving your event and animation supports the story.

- **Falling Action:** Where did your event go? There is nothing on the screen that shows you where your event went or how to return to it if you want to edit it. You hope you saved and entered everything in correctly. This can be a crisis or a cliffhanger, depending on how it plays out in real life.

- **End:** If you're like me, you probably added the event to the wrong day, oops. Or you take a leap of faith and assume that you entered it correctly and will be at coffee on time.

Here's how the same core task works in another iOS calendaring app, Fantastical:

- **Exposition:** You're having lunch with Elon in Palo Alto tomorrow. How exciting. You definitely don't want to miss this.

- **Inciting Incident:** You tap the "Add Event" button to add the event to your calendar (see Figure 5.11).

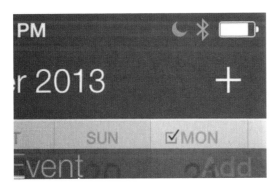

FIGURE 5.11
The inciting incident or call to action of adding a calendar event to Fantastical: a button that lets you add a new event.

- **Rising Action:** You start typing "Lunch with Elon in Palo Alto" (see Figure 5.12), and the screen starts to fill in your location information on a timeline visualization that shows you when and where lunch is (see Figure 5.13). How smart—it assumes that lunch is at noon. When you're done adding your event, you tap the *Add* button (see Figure 5.14).

- **Crisis:** There is none because...

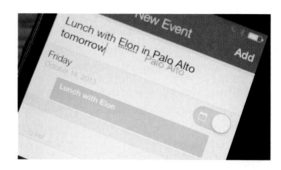

FIGURE 5.12
Rising action for Fantastical: the text you type animates to indicate that something is happening.

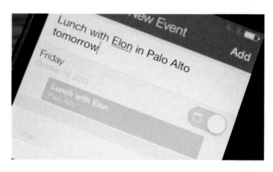

FIGURE 5.13
Rising action for Fantastical: you see that the app automatically parsed your event type and location for noon in Palo Alto.

FIGURE 5.14
Rising action for Fantastical: saving your calendar event.

- **Climax/Resolution:** The event shrinks and animates onto the correct day in the calendar (see Figure 5.15), and at the end of that animation, it glows red (see Figure 5.16) before the red fades away (see Figure 5.17). It's such a subtle, fast animation, but it's helpful to see feedback after you complete a task. Feedback in this case *communicates* that your event was added and added to the correct day. The animation also supports the mental model of events being placed on a calendar. "Fast and friendly" is how Fantastical's developers describe the app, and this climax embodies that description. They say it in their marketing materials, and you experience it in the micro-interaction.

FIGURE 5.15
Fantastical: crisis averted. Animation communicates system feedback that your event has been saved.

FIGURE 5.16
Fantastical: crisis averted. The rest of the animation communicates that your event has been saved on the correct date: tomorrow, October 18.

FIGURE 5.17
Fantastical: All done. The end...for now.

- **Falling Action/End:** You have no doubt that this calendar event was added, not just to your calendar, but also to the correct day. Task complete. You're glad you downloaded this app to use instead of the Calendar app, which is so *not* smart.

Something as simple as a calendar app, with a core task that you want users to repeat and repeat, can benefit from this structure. Otherwise, your core task flow falls flat, just like the built-in iOS Calendar app did. Is Apple in trouble? Not at all. Unless calendaring becomes core to their business and engagement strategies.

When software is your service and your primary means for acquiring and retaining customers, you need to make sure that everything, no matter how small, reads like a story. Contrast something seemingly trivial like iOS's Calendar app to the unboxing experience for the iPhone. The latter absolutely reads like a story, from the moment you rip off the plastic wrap to the minute you turn on your phone... and then continues into the setup UI. The iPhone is core to Apple's strategy. The iPhone is built on story. Many stories. When it matters, you should plot even the smallest and seemingly trivial core tasks story-first.

For even the smallest interactions, good design incorporates affordances, minimizes steps to completion, and gives users feedback. Story does all of those things, as well.

A good story is good design.

Case Study: FitCounter

In the case of the start-up, *FitCounter*, as we ideated, tested, and gathered qualitative and quantitative feedback from existing and potential customers, we started to feel like we had a product-market fit with the *concept* of what the product and service could be. We also successfully envisioned and engineered *origin stories* that helped visitors find the product, want to use it, and start on their journey of becoming more fit. But having that concept and the first contact was not enough. We needed to create a product that people actually used. Eventually, we also had to see if people would pay to use it, but our hunch was that they needed to try it out first before they could decide whether or not to pay for the upgrade.

In order to see if we had a viable product, we needed to envision, assess, test, and build a minimum representation of a product and service that delivered on this promise of helping people get fit, stay fit, and help others get and stay fit. Our front door was inviting enough for potential customers to want to sign up, but now we needed to get them to actually sign up, do something, and *experience* value.

The Problem: Broken Funnel... and No Engagement

Much like Twitter, FitCounter's previous sign-up flow was only a few steps. The team designed it that way to be as fast and frictionless as possible. The sign-up flow even had a progress bar so that new users knew what steps there were in the process and how much longer they had to go.

Despite the flow being straightforward, simple, and easy to use, the funnel completion rates were rather low overall, as well as being low from one step to the next (see Figure 5.18). Not many new users who started the flow completed the sign-up process. And of those few who did sign up, even fewer ever tried using the product afterward. Why was this happening?

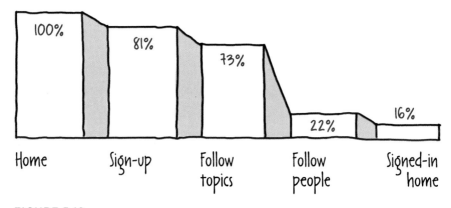

FIGURE 5.18

FitCounter's problematic on-boarding funnel shows the drop-off from step to step and overall.

The team had been iterating on this flow for quite some time, and they only seemed to improve things by a couple of percentage points each time they tried.

When you map out the previous sign-up funnel as a story, it looks more like what you see in Figure 5.19, which is a cliffhanger.

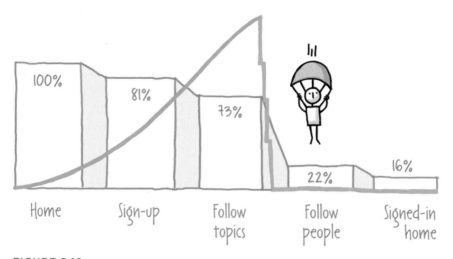

FIGURE 5.19
The cliffhanger of this funnel is clear when visualizing it as a story. A majority of visitors who started the funnel were signing up, but not continuing after the next step.

Data Tells a Story

Both analytics funnels and stories describe a series of steps that users take over the course of a set period of time. In fact, as many data scientists and product people will tell you, data tells a story, and it's our job to look at data within a narrative structure to piece together, extrapolate, troubleshoot, and optimize that story.

In the case of FitCounter, our gut-check analysis and further in-person testing with potential users uncovered that the reason our analytics showed a broken funnel with drop-off at key points was because people experienced a story that read something like this:

- **Exposition:** The potential user is interested in getting fit or training others.

- **Inciting Incident:** She sees the "start training" button and gets started.

- **Rising Action:**

  - She enters her username and password. (A tiny percentage of people would drop off here, but most completed this step.)

  - She's asked to "follow" some topics, like running and basketball. She's not really sure what this means or what she gets out of doing this. She wants to train for a marathon, not follow things. (This is where the first drop-off happened.)

- **Crisis:** This is where the cliffhanger happens. She's asked to "follow" friends. She has to enter sensitive Gmail or Facebook log-in credentials to do this, which she doesn't like to do unless she completely trusts the product or service and sees value in following her friends. Why would she follow them in this case? To see how they're training? She's not sure she totally understands what she's getting into, and at this point, has spent so much brain energy on this step that she's just going to bail on this sign-up flow.

- **Climax/Resolution:** If she does continue on to the next step, there would be no climax.

- **Falling Action:** Eh. There is no takeaway or value to having gotten this far.

- **End:** If she does complete the sign-up flow, she ends up home. She'd be able to search for videos now or browse what's new and popular. Searching and browsing is a lot of work for someone who can't even remember why they're there in the first place. Hmmm…in reality, if she got this far, *maybe* she would click on something and interact with the product. The data told us that this was unlikely. In the end, she didn't meet her goal of getting fit, and the business doesn't meet its goal of engaging a new user.

Why was it so important for FitCounter to get people to complete this flow during their first session? Couldn't the business employ the marketing team to get new users to come back later with a fancy email or promotion? In this case, marketing tried that. For months. It barely worked.

With FitCounter, as with most products and services, the first session is your best and often only chance to engage new users. Once you grab them the first time and get them to see the value in using your product or service, it's easier to get them to return in the future. While I anecdotally knew this to be true with consumer-facing products and services, I also saw it in our data.

Those superfans I told you about earlier rarely became superfans without using the product within their first session. In fact, we found a sweet spot: most of our superfans performed at least three actions within their first session. These actions were things like watching or sharing videos, creating playlists, and adding videos to lists. These were high-quality interactions and didn't include other things you might do on a website or app, such as search, browse, or generally click around.

With all of our quantitative data in hand, we set out to fix our broken usage flow. It all, as you can imagine, started with some (more) data...oh, and a story. Of course.

The Plan

At this point, our goals with this project were two-fold:

- To get new users to complete the sign-up flow.
- To acquire more "high-quality" users who were more likely to return and use the product over time.

As you can see, getting people to pay to upgrade to premium wasn't in our immediate strategic roadmap or plan. We needed to get this product operational and making sense before we could figure out how to monetize. We did, however, feel confident that our strategy was headed in the right direction because the stories we were designing and planning were ones that we extrapolated from actual paying customers who loved the product. We had also been testing our concept and origin stories and knew that we were on the right track, because when we weren't, we maneuvered and adapted to get back on track. So what, in this case, did the data tell us that we should do to transform this story of use from a cliffhanger, with drop-off at the crisis moment, to a more complete and successful story?

Getting to "Why"

While our quantitative analytics told us a "what" (that people were dropping off during our sign-up funnel), it couldn't tell us the "why." To better answer that question, we used story structure to figure out why people might drop off when they dropped off. Doing so helped us better localize, diagnose, and troubleshoot the problem. Using narrative structure as our guide, we outlined a set of hypotheses that could explain why there was this cliffhanger.

For example, if people dropped off when we asked them to find their friends, did people not want to trust a new service with their login credentials? Or did they not want to add their friends? Was training not social? We thought it was. To figure this out better, once we had a better idea of what our questions were, we talked to existing and potential customers first about our sign-up flow and then about how they trained (for example, alone or with others). We were pretty sure training was social, so we just needed to figure out why this step was a hurdle.

What we found with our sign-up flow was similar to what we expected. Potential users didn't want to follow friends because of trust, but more so because it broke their mental model of how they could use this product. "Start training" was a strong call to action that resonated with potential users. In contrast, "follow friends," was not. Even something as seemingly minute as microcopy has to fit a user's mental model of what the narrative structure is. Furthermore, they didn't always think of training as social. There were a plethora of factors that played into whether or not they trained alone or with others.

What we found were two distinct behaviors: people tend to train alone half the time and with others half the time. Training alone or with others depended on a series of factors:

- Activity (team versus solitary sport, for example)
- Time (during the week versus weekend, for example)
- Location (gym versus home, for example)
- Goals (planning to run a 5k versus looking to lose pounds, for example).

This was too complex of a math equation for potential users to do when thinking about whether or not they wanted to "follow" people. Frankly, it was more math than anyone should have to do when signing up for something. That said, after our customer interviews, we were convinced of the value of keeping the product social and giving people the opportunity to train with others early on. Yes, the business wanted new users to invite their friends so that the product could acquire new users. And, yes, I could have convinced the business to remove this step in the sign-up process so that we could remove the crisis and more successfully convert new users. However, when people behave in a certain way 50% of the time, you typically want to build a product that helps them continue to behave that way, especially if it can help the business grow its user base.

So instead of removing this troublesome cliffhanger-inducing step in the sign-up flow, we did what any good filmmaker or screenwriter would do: we used that crisis to our advantage and built a story *with* tension and conflict. A story that we hoped would be more compelling than what we had.

The Story

In order to determine how our new sign-up flow would work, we first mapped it out onto a narrative arc. Our lead designer and engineer wanted to jump straight into screen UI sketches and flow charts and our CEO wanted to see a fully clickable prototype yesterday, but we started the way I always make teams and students start: with a story diagram. As a team, we mapped out a redesigned sign-up flow on a whiteboard as a hypothesis, brick by brick (see Figure 5.20).

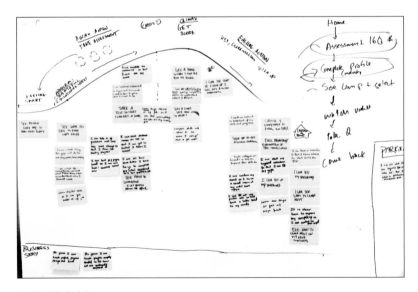

FIGURE 5.20

A story map from a similar project with the storyline on top and requirements below.

This was the story, we posited, that a new user and potential customer should have during her first session with our product (see Figure 5.21). As you can see, we tried to keep it much the same as before so that we could localize and troubleshoot what parts were or weren't working.

- **Exposition:** She's interested in getting fit or training others. (Same as before.)

- **Inciting Incident:** She sees the "start training" button and gets started. (Same as before.)

- **Rising Action:**

  - She enters her username and password. (This step performed surprisingly great, so we kept it.)

  - Build a training plan. Instead of "following" topics, she answers a series of questions so that the system can build her a customized training plan. Many questions—ultimately extending the on-boarding flow by 15 screens. 15! There is a method to this madness. Even though there are now many more questions, they get more engaging, and more relevant, question by question, screen by screen. The questions start broad and get more focused as they progress, feeling more and more relevant and personal. Designing the questionnaire for rising action prevents what could be two crises: boredom and lack of value.

- **Crisis:** One of the last questions she answers is whether or not she wants to use this training plan to train with or help train anyone else. If so, she can add them to the plan right then and there. And if not, no problem—she can skip this step and always add people later.

- **Climax/Resolution:** She gets a personalized training plan. This is also the point at which we want her to experience the value of her new training plan. She sees a graph of what her progress will look like if she sticks with the training plan she just got.

- **Falling Action:** Then what? What happens after she gets her plan and sees how she might progress if she uses FitCounter? This story isn't complete unless she actually starts training. So...

- **End:** She's home. Now she can start training. This initially involves watching a video, doing a quick exercise, and logging the results. She gets a taste of what it's like to be asked to do something, to do it, and to get feedback in the on-boarding flow and now she can do it with her body and not just a click of the mouse. Instead of *saying* how many sit-ups she can do by answering a questionnaire, she watches a short video that shows her how to best do sit-ups, she does the exercise, and she logs her results. While humanly impossible to fully meet her goal of getting fit in one session, completing the story with this ending gets her that much closer to *feeling* like she will eventually meet her goal. Our hope was that this ending would function as a teaser for her next story with the product, when she continued to train. We wanted this story to be part of a string of stories, also known as a *serial story*, which continued and got better over time.

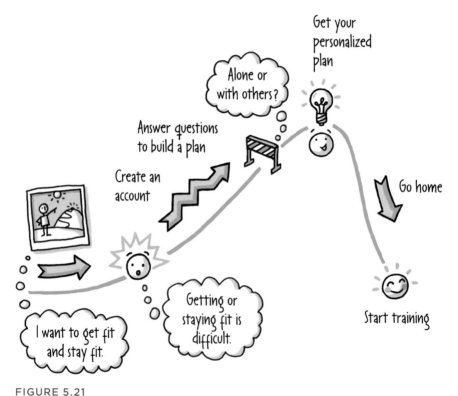

FIGURE 5.21

The story of what we wanted new users to experience in their first session with FitCounter.

Once we plotted out this usage story, we ran a series of planning sessions to brainstorm and prioritize requirements, as well as plan a strategic roadmap and project plan. After we had our requirements fleshed out, we then sketched out screens, comics, storyboards, and even role-played the flow internally and in person with potential customers. We did those activities to ideate, prototype, and test everything every step of the way so that we could minimize our risk and know if and when we were on the right path.

We were quite proud of our newly crafted narrative sign-up flow. But before we could celebrate, we had to see how it performed.

The Results

On this project and every project since, we tested *everything.* We tested our concept story, origin story, and everything that came after and in between. While we were very confident about all of the work we did before we conceived of our new usage story for the sign-up flow, we still tested that. Constantly. We knew that we were on the right path during design and in-person testing because at the right point in the flow, we started getting reactions that sounded something like: "Oh, cool. I see how this could be useful."

Once we heard that from the third, fourth, and then fifth person during our in-person tests, we started to feel like we had an MVP that we were not only learning from, but also learning *good* things from. During our concept-testing phase, it seemed like we had a product that people might want to use. Our origin story phase and subsequent testing told us that the data supported that story. And now, with a usage story, we actually had a product that people not only *could* use, but *wanted* to use. Lots.

As planned, that reaction came during our in-person tests, unprompted, near the end of the flow, right after people received their training plan. What we didn't expect was that once people got the plan and went to their new home screen, they started to tap and click around. A lot. And they kept commenting on how they were surprised to learn something new. And they would not only watch videos, but then *do* things with them, like share them or add and remove them from plans.

But this was all in person. What about when we launched the new sign-up flow and accompanying product. This new thing that existed behind the front door. The redesign we all dreaded to do, but that had to be done.

I wish I could say that something went wrong. This would be a great time to insert a crisis moment into this story to keep you on the edge of your seat.

But the relaunch was a success.

The story resonated not just with our in-person testers, but also with a broader audience. So much so that the new sign-up flow now had almost double the completion rate of new users. This was amazing, and it was a number that we could and would improve on with further iterations down the line. Plus, we almost doubled our rate of new user engagement. We hoped that by creating a sign-up flow that functioned like a story, the result would be more engagement among new users, and it worked. We not only had a product that helped users meet their goals, but it also helped the business meet its goals of engaging new users. What we didn't expect to happen so soon was the side effect of this increased, high-quality engagement: these new users were more likely to pay to use the product. Ten times more likely.

We were ecstatic with the results. For now.

A business cannot survive on first-time use and engagement alone. While we were proud of the product we built and the results it was getting, this was just one usage story: the first-time usage story. What about the rest? What might be the next inciting incident to kick off a new story? What would be the next beginning, middle, and end? Then what? What if someone did not return? Cliffhangers can happen during a flow that lasts a few minutes or over a period of days, months, or years. Over time, we developed stories big and small, one-offs and serials, improving the story for both customers and the business. Since we started building story-first, FitCounter has tripled in size and tripled its valuation. It is now a profitable business and recently closed yet another successful round of financing so that it can continue this growth.

CHAPTER 6

Finding and Mapping Your Story

—Alfred Hitchcock

Where do stories come from? Stories are not something that you make up (although there is a place for that, as you'll see later in this chapter). Stories for products come from data—data that your business already has or data that you seek through qualitative and quantitative research. Mapping stories isn't a creative endeavor—it's a strategic business tool and activity tied not just to real data but also to real results, metrics, and KPIs. Mapping stories helps you figure out what is and what can be for your product, your customers, and your business.

But how do you see stories in your data? The short answer is that the stories are already there. They are just waiting for you to uncover and do something with them. In order to learn how to craft a story, you must first learn to *see* stories. Once you start looking for them, you will begin to see them everywhere. The question you will begin to posit isn't *what's the story?* But, rather: *Is this story any good?*

For example, think about your competitor's product that is doing really well. You'll find on closer examination that concept, origin, and usage stories probably run through it. The relationship you have with your favorite product or brand? Also a story—the story of why you like it, why you continue to use it, and why you recommend it to others. Stories can be found in how your customers use your products (or competitor's products), how they talk about your product and the problems it helps them solve, and how they experience using your product or engaging with your brand. Looking for stories is the first step you take toward not just finding but building stories into the things (products or whatever) you put out into the world.

Listen

Listening for stories usually takes the form of getting out of the building and talking to your customers or those people you hope will eventually be your customers. Who are they? What is good in their world? What problems do they have? What might call them to action or solve their problems? How do they do things? What makes them tick? After you start talking and listening to your customers, you will start to hear patterns and themes emerge that you can parse together with structure. These are their stories—on the conceptual level and the behavioral level. And they become the storyline of your product.

Your brain already uses story structure to make sense of other people's words and actions. Now is your chance to get deliberate about it. Map your customers' stories as you hear them, plot point by plot point. Make sense of their stories. Find the patterns. Find the gaps. See where you can improve their stories. More specifically, see how your product can support their stories.

When doing research, you can parse, make sense of, and map out the stories that you hear on a wall (see Figure 6.1) or on a computer, if your team is remote (see Figure 6.2). Or you can map out your stories on the fly as little squiggles as they emerge during and after you talk to customers (see Figure 6.3). At their most powerful, storylines are strategic tools for uncovering patterns, as well as opportunities.

FIGURE 6.1

If you map out findings from user research on a wall with Post-it notes, you can use plot points to organize your thoughts and insights.

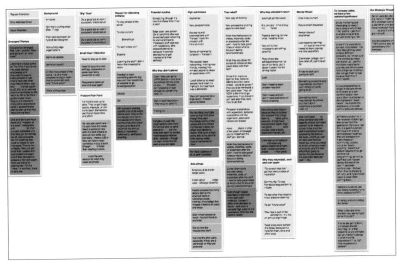

FIGURE 6.2

If your team isn't co-located, you can use a spreadsheet or online collaboration tool like BoardThing to map out insights from research.

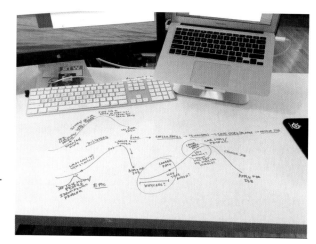

FIGURE 6.3
These storylines visual-
ize insights gathered
while doing remote
research.

Use the Smile Test

When I map out storylines with teams, I often see one key smiley face
scribbled on it, usually in the same place: the climax (see Figure 6.4).
It's what you hope will be the high point of an experience with a
feature, flow, or overall product. If you uncover your storylines from
existing customers or by mapping out real behavior that real people
do, this smile might be something that you observed doing field
research, usability testing, or customer interviews. Once you've got
a working prototype at any level of fidelity, you can test out your
story to ensure that your climax is something that really makes your
customers smile.

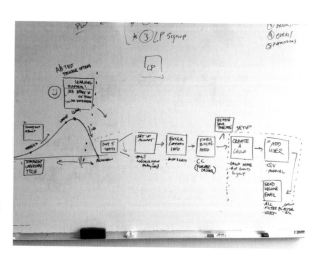

FIGURE 6.4
A storymapping exer-
cise that helped the
team imagine and test
why people would use
the product in the first
place. This became the
business's origin story
for its core customers.

Smiling at FitCounter

At FitCounter, we designed the on-boarding flow to teach customers how the system worked and give them a taste of the value—something they could use to train in bite-sized chunks, on their own time or with others. Our hope was that if they *felt* what it was like to use the service, they would interact with it some more during their first session. Our data told us that if customers took five actions in their first session (either watching videos or logging data or a combination of both), they were more likely to convert to a long-term and paying customer. But we didn't want new users to just do five things; we wanted them also to experience the value of doing those things. We placed little smileys at the climaxes of our story diagrams to remind us during the ideation and development process that we needed to make the customers feel like they had just spent the last five minutes of their lives wisely and gotten something out of it.

We prototyped these stories as quickly as possible in the lowest fidelity we could think of—conversations—and then worked our way up to paper, clickable lo-fi prototypes, and eventually code. Each time we would graduate to a new level of fidelity, we tested our prototype in person with existing customers and people who could be future customers. What we found during each round of testing was that we would see customers grin at the right point, validating our hunch that we were not just solving a problem but also providing value. Sometimes, the smile was accompanied by a statement like, "Oh, that was neat" or "Ah, I get it."

Before we had redesigned our product, we got comments like "I don't get it," or worse yet, long pauses, grimaces, and polite yet insincere "hmm...interesting..." as people played around with the product. Afterward, smiles trumped words—we not only heard we were onto something, but we could *see* it.

Measure

Stories can be found in qualitative data you gather from your customers as well as the quantitative data that you collect from your business intelligence and analytics tools. Consider funnel analytics that you find in typical software like Google Analytics. When do people drop off? Those are your cliffhangers.

When you see a cliffhanger in your analytics, you see the *what*: data. Stories tell you the *why*. Imagine that you are analyzing a checkout flow, and you see that there is a significant drop-off at step 3. Let's say, step 3 is when your business asks users to enter their payment information. There might be many reasons why users drop off at this point. Maybe they are new users who don't trust your business enough to enter their credit card details. Maybe they have questions and would rather chat on the phone with a sales representative. Maybe they live in a foreign country or are underage and don't have access to a credit card. Maybe the checkout flow is confusing, and they thought they submitted their order, when they never actually did. I've worked on products that have had all of these scenarios as the *why* behind why checkout conversion wasn't as high as the business wanted it to be. Each of the scenarios has a simple fix because each of the scenarios is a simple story. For every cliffhanger and crisis moment, there's a climax that can resolve that problem.

> Someone wants to do something. Something gets in their way. Something can help them overcome this obstacle. They get what they want. The end.

Next time you uncover the *what* while looking at data and analytics, ask yourself *why*? Sometimes, you'll find that you can whip up a quick story map as a hypothesis to answer the why (which you then have to test). Plot point by plot point. What might seem obvious or mysterious is usually not what you think or is completely solvable once you map it out as a story. Often, you won't be able to answer the why on your own, and you'll have to leave the building and talk to a few customers or watch them using your product. What you will uncover will be their story, plot points, cliffhanger, and all. And what you'll craft as your solution or fix will be a better story—one that has a beginning, middle, *and* an end.

Case Study: SmallLoans—a Cliffhanger

While you can find stories in your funnel analytics, you can also find them in complex data sets and business intelligence data. Take, for example, a mobile micro-lending start-up that a colleague of mine used to work with. I'll call it SmallLoans to protect the innocent. SmallLoans provides micro-loans to people in developing countries via a simple mobile Web interface. Because bandwidth is low and technology is old in developing countries, the service needs to function seamlessly, with as few errors and blips as possible. Phone calls to customer service and jumping on a desktop computer to troubleshoot are not options for these customers when something goes wrong. If anything goes wrong during the SMS-based loan application process, it taxes the customer service team, or worse yet, compels potential customers to go elsewhere for their loans.

Early in the life of this start-up, things were going very wrong: it was getting inundated with customer service requests and needed to figure out how to minimize the volume. Potential customers were also slamming customer service with SMS texts, which was taxing the system and the team. It's difficult enough to have complex text-based conversations with your friends and family—imagine trying to conduct customer service triage in 160 characters or less.

People were signing up for the service because they knew that they could use it to secure a small loan quickly. But after a day, they simply wanted to know the status of their loan. Not knowing is hard—especially if your livelihood depends on it. SmallLoans tried to figure out ways to ping new applicants after they had signed up to pre-empt issues, or ways to train their customer team to handle these requests better with canned responses.

On the more complex side, they started to look at their data to see if they were taking too long to approve loans and how they might fix that. The solution they arrived at was much simpler than all of these solutions, and it tapped into the concept story of the service at its core.

SmallLoans prided itself on a fast turnaround time for loan approval—maybe a few days, maximum. But when the customer service requests multiplied, they looked at their data and saw that their turnaround time for loans was 36 hours.

That was fast. Much faster than the team had thought.

They knew their technology and process were good, but not that good. Think about it—imagine if you knew that you could be approved for a loan within 36 hours. No back-and-forth with application forms. No phone calls to loan officers. No waiting weeks or months.

36 hours.

That's the high point of a pretty good experience—it's not just fast, but 36 hours fast—it was their *climax*.

That climax is part of the story that SmallLoans eventually communicated, not during the process or after an application was submitted, but *before* someone applied for a loan.

SmallLoans took what was a crisis in their customer's origin stories— a cliffhanger if they decided to go with a competitor instead—and used that potential conflict to make the exposition of the customer journey that much more compelling. They started to let potential customers know up-front, right before step 1, that the loan process was not just fast but guaranteed to take no more than 36 hours.

With this little change in how they crafted their origin story for new customers, they ended up decreasing customer service complaints and increasing conversion and funnel completion significantly. SmallLoans not only crafted a more structurally sound story for the sake of making things better, but it also saw measurable results.

Sometimes, all you need to do is find the story in your data and communicate and amplify it for your customers.

Innovate: What If?

"The most interesting situations can usually be expressed as a *What-if* question: What if vampires invaded a small New England village? What if a policeman in a remote Nevada town went berserk and started killing everyone in sight?"

—Stephen King,
On Writing: A Memoir of the Craft

Storylines should be data-driven, but they can also be magical, fantastical, or simply delightful.

When interviewing users, I often ask this question: "What if this product worked like magic?" At first, people are startled. Then they're confused. "Magic," they ask? "What do you mean?" Once they get over the mental hurdle of essentially being asked to design the product for me, they eventually have fun with their answers. "It would just know what I want!" they might say. "It would order dinner for me without me having to choose what to eat." "It would tell me what's good so I don't have to seek it out." "It would tell me what I need to learn and when I need to learn it."

Imagining *What If* with FitCounter

With FitCounter, the answer to the "what if" question was something like this: "It would give me a training plan based on my level of fitness, my short- and long-term goals, the time I have to devote to training, and the times of the day and week I prefer to work out." Most importantly, however, the answer came down to "It will make me work out and stick with it over time."

When we redesigned FitCounter, apps like RunKeeper allowed you to track your workouts and set goals. Apps like Couch to 5K gave you more detailed plans, but started you at ground zero, even if you could already run a mile without stopping, and forced you to continue onto 5K, even if your goal was just to run two miles. In-person trainers could personalize a plan for you, but they were expensive and required travel. *What if* we could build a product that asked you a few questions, built a personalized plan for you, and checked in to make sure that you were staying on track. And if training with others helped you stay on track, you could use FitCounter with your friends.

While this seemed like a tall order, this was something that our engineering team could figure out and build by crafting an on-boarding flow that collected key data from people and then built them a training plan based on that data we collected. In fact, this is something that fitness trainers— the human kind—do every day with new clients. They talk to clients, find out what their goals are, how much time they have to devote to training, what types of exercises they do and don't enjoy, and build a plan accordingly. Then, over time, the trainer works with people to make sure that they stick with the plan.

What if was our question. Mapping the story out gave us our answer. We heard the story from our customers. We saw stories while looking at the competitive landscape. We vetted them internally as we collected team members' *what-if* scenarios they had been imagining for years. Then, once we had a story—many stories, we *used* them. In the next chapter, you'll see how.

Borrow: Stories as Proofs of Concept

Sometimes, you or your business will invent a new product or feature based on a bout of inspiration. *What if,* you ask, *I could order a taxi at the tap of a phone screen?* That's a story, yup. Often, however, you see that filmmakers and science fiction writers have done all of the creative work for you. In that case, you can see storylines that flow through fictional products that excite you. Then you can map them out to assess their strengths, weaknesses, and product-market fit. Doing so will not only help you build the right product but also build it for the right market.

1989...and 2001: An iPad Odyssey

For example, recently I stayed in London in a different time zone than my family. At the end of a long day working, I returned back to my apartment, bored and lonely. I took my iPad with me so that I could catch up on *Star Trek the Next Generation,* while my partner, Erica, did the same, five hours after me back in Brooklyn, where we lived. I was watching *Star Trek* partly for fun, but more so for research. I wanted to see how the characters used wearable technology, like eyewear, in this fictionalized future. I was working on a project about the future of devices like Google Glass, and hoped I would find some answers by doing one of my favorite things: watching TV.

The episode I watched, which originally aired in 1989, centered on a little boy who lost his parents and was alone on the Starfleet Enterprise. In one scene, he missed his mom and pulled out this tablet device that looked an awful lot like an iPad so that he could watch old home movies. This was the only way he had to connect with her—so wherever he was, his mom was with him. At this moment in the scene, I took a photo (see Figure 6.5).

What you see in this photo is me using my iPad to watch a moving picture of a boy using an iPad-like device to watch a moving picture. The use cases were similar, and the stories that flowed through those use cases were solid: someone wants to connect or communicate with someone or something far away. With the iPad (the product), the person can connect (the story).

FIGURE 6.5
A photograph of
me using my iPad to
watch a character
on *Star Trek* watch
a home movie on a
touchscreen device
that looks uncannily
like an iPad.

Tablets like the iPad solve problems and move action forward in storylines both in science fiction and real life. And people who build technology are no strangers to these props and images from science fiction. Steve Jobs even "borrowed" the name of the iPad from the NewsPad featured in *2001: A Space Odyssey* (see Figure 6.6). In that story, the characters used this device to communicate with home using newsfeeds and video. It was a pretty powerful device. But more so, the storyline behind using the device as a prop within the film was sound. As any filmmaker will tell you, everything that happens in a film *must* move the story forward or it gets cut.

FIGURE 6.6
A still from *2001: A Space Odyssey* showing someone using a device in much the same way someone would use an iPad today.

2014: A Google Glass Odyssey

Contrast the stories of something like the iPad with a proof of concept like Google Glass, which is essentially a computer that you wear on your face. I have spent probably 100 hours over the past couple of years trying to figure out if, why, or how a device like Glass and apps built on that platform could be viable. At first, I was convinced that the device was a dud and that consumers would never need it or pay money for it. Actually, this has turned out to be true. Google discontinued the program and is no longer producing or marketing Glass as a consumer device. However, when you look at sci-fi, you see a different story. Wearable technology, like Glass, has a product-market fit. It's just not a mass-market-fit.

There are three storylines to be exact (see Figure 6.7):

- Someone needs her hands free so that she can do her job (like fight aliens, paramilitary, surgery, or police work—e.g., *They Live, Terminator, Iron Man, Star Trek: The Next Generation, RoboCop*).

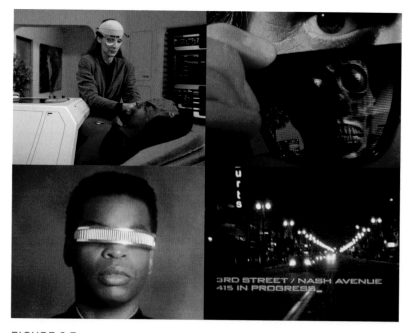

FIGURE 6.7

Clockwise from top, left, characters use wearable technology to: perform a medical procedure (*Star Trek: The Next Generation*), see concealed alien communication (*They Live*), navigate (*RoboCop*), and see (*Star Trek: The Next Generation*).

- Someone is moving very fast and needs his hands free so that he can use his hands for other things…and do his job (*Iron Man, RoboCop*).

- Someone has a physical impairment and this technology gives him an ability to see and hear (*Star Trek: The Next Generation, RoboCop*).

What looking at these storylines uncovers is that a device like Glass is not a mass-market consumer device. It is best suited for work or being human when you have some kind of impairment. Furthermore, this type of technology already exists in these contexts to help people get stuff done and be human. For example, while I sit here writing, I am wearing glasses—the oldest form of wearable face technology out there. And the military, police, soldiers, firefighters, and doctors have been wearing technology on their bodies for years. This takes the form of helmet-mounted displays and lighting mechanisms, body-mounted video cameras, and surgical loupes (see Figure 6.8). When you've got a job to do and need your hands free to do it, technology can help. *And that's a structurally sound story.*

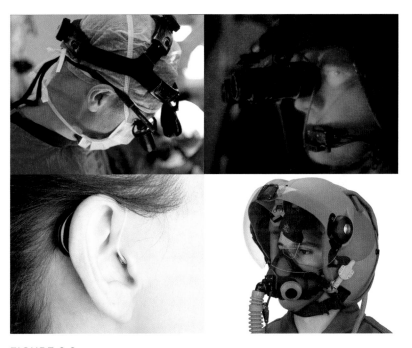

FIGURE 6.8
Clockwise from top, left, people in real life use wearable technology to perform a medical procedure (surgical loupes), see at night (infrared goggles), navigate (helmet-mounted heads-up display or *HUD*), and hear (hearing aid).

If you're working on an entirely new product or feature, you don't have to get stuck imagining storylines from scratch. Learn from what came before you. Hollywood produces some of the most expensive proofs of concepts and prototypes in the world. They're called *props*. And each prop is a key player in a structurally sound storyline. See how filmmakers use these products to support the story. Glean meaning from how characters use them to be heroic and save the day.

When you're finding stories in the real world, whether they are in movies or your customers' lives, remember to first see and hear your stories; then build them. Think like a storyteller and learn from and borrow from what other creative people—filmmakers and even your customers—out there are doing before you chart your course anew. Or if you must, chart your course anew. You might call this paving the cow path, building empathy, or being creative. I call this *storymapping*.

CHAPTER 7

Using Your Story

"As for the story, whether the poet takes it ready made or constructs it for himself, he should first sketch its general outline, and then fill in the episodes and amplify in detail."

—Aristotle,
Poetics

When I first started to teach people how to map out stories for product and service design and development, I gave them the choice: use your story as a loose guide, or plot methodically onto a narrative arc diagram. I don't like to dictate process, nor do you want to be told exactly how to do your work. That said, I will tell you this—at least while you're starting out using this technique, map *everything*. And do so visually. On a squiggly narrative arc. Then as you explore your stories in different mediums and fidelities, expect that story to change. Your story maps are more like guides than skeletons—they are loose paths for how you *intend* for people to experience something. As such, they can and should evolve as you explore, as you plan, as you build, and as people interact with what you put out into the world.

I map stories out on a narrative arc because it's my preferred (read: *simple*) story diagram of choice and the one that most of my clients and team members grasp most quickly. The narrative arc also models how humans interact with products as moments in time with its linearity and peak near the end. Mapping concepts and flows onto a narrative arc helps you see the flow of ideas and interactions as your users might emotionally experience them.

If you have a background in film or creative writing, you might have another type of diagram you prefer. If so, use that. There are almost as many permutations of narrative architectural diagrams as there are stories. Just remember: the key to all of this is making sure that you have a story that flows through everything you build, not what type of diagram you use.

Squiggly lines on a piece of paper or a whiteboard are cheap. Not having a storyline or losing your thread is expensive. Once you've got that squiggly yarn, you can and should thread, prototype, test, and build it into many of the things that you already do: diagram, sketch, write, pitch, analyze, communicate, prototype, build, test… everything. You can explore and evolve your stories on your own or with a group—on the fly or as an organized workshop.

Illustrate Your Story with Strategic Tools

One way to bring your storylines to life is to explore them visually. It might start as a squiggle and then grow into something much bigger and more complex. Activities like diagramming, storyboarding, or journey mapping are strategic tools you might already have in your arsenal. They are each that much more powerful when you use them to uncover, support, or convey your story with your team, stakeholders, or clients.

Diagram

Flow charts and diagrams are visual stories. You can start off a flow-charting session on a whiteboard by inserting your storyline so that you remember what the story is.

If you do so, as you should do with all flow charts, remember to build around your "happy path" or ideal storyline of how you want someone to experience something. Then, if need be, consider branching paths, edge cases, and alternate scenarios. Systems are branching, and time and experience are linear. Always plan for how you want things to work out and then deviate accordingly. Doing so will not only help you retain your sanity while creating diagrams, but will also remind you how to help people navigate what otherwise might be really complex spaces and systems.

Storyboard

You can also turn your storylines into story*boards* or comics, a visual representation of your story that is laid out as panels in a grid.[1] You can use storyboards to visualize the big picture of how someone thinks about or uses your product (see Figure 7.1) or how to get more detailed and map out specific steps in a flow or interface.

It's best to keep your storyboards as short as possible—ideally, no more than nine panels. Doing so helps you ensure that your story is there, because it's easy to lose your storyline when your scope gets too big or your details too plentiful. Keeping your storyline to nine panels also helps you remember to get to the point—plot points. When illustrating a storyboard, you'll want to make sure that your inciting incident happens as quickly as possible and that your climax

---

1 For more on creating storyboards and comics, see Cheng, Kevin. *See What I Mean: How to Use Comics to Communicate Ideas.* Brooklyn, N.Y.: Rosenfeld Media, 2012.

or resolution leads to a very swift ending. This means that the inciting incident and climax should happen on the second and seventh panel, respectively (see Figure 7.2). Nothing is worse than a beginning or an end that drags on longer than it needs to, whether it is in the telling of a story or experiencing it. Storyboards, just like your squiggles, are essentially a prototype of an experience. Visualizing in this way helps you make sure that your scope and pacing are tight.

FIGURE 7.1
Panels from a sample storyboard. (Courtesy of *See What I Mean* by Kevin Cheng)

FIGURE 7.2
Storyboarding in nine panels helps you see the key inciting incident and climax plot points, as well as keep your scope manageable.

Strategic Storymapping

> "There's always going to be an entertainment factor that goes into what you're designing. [But] no matter what, you're designing to support the story."
>
> —Tim Flattery,
> "Future Consultant," *Back To The Future Part II*[2]

Storymapping is an excellent way to visually map out a customer or user journey. It will also help you quickly, effectively, and collaboratively assess all the things you need to support the story and make that journey successful for your user and your business. You might already do similar mapping exercises: *journey mapping, customer journey mapping, user journey mapping, experience mapping*, or *agile user storymapping*. I'm sure I'm missing a few names here. No matter what you call it or how you do it, if you want to engage your audience, start with a story map as your framework. Then build on (and under) that map to determine what you need to support your story for your user and your business. Literally, map out the story first and then flesh it out and fill it all in afterward with Post-it notes or cards (see Figure 7.3).

FIGURE 7.3
Mapping exercises start with a storyline (left) and can then build into something more complex (right).

---

2 Trendacosta, Katharine. "Brand New Concept Art of *Back to the Future Part II's* 2015 Technology." io9. http://io9.com/brand-new-concept-art-of-back-to-the-future-part-iis-20-1678913570

Here are some ways that you can use story maps to solve different problems:

- **Gap analysis** (see Figure 7.4): You might want to figure out how to get more customers to sign up or get through a flow. Mapping the story as a gap analysis exercise is a great way to visualize the gap between the current state of a journey and the future state that you will build.

FIGURE 7.4

Gap analysis: Using story to visualize the gap between a cliffhanger (left) and a flat story (right) and complete stories that meet user and business goals.

- **Behavior analysis:** Sometimes, you want to know what the story is for different types of users. In this case, you might map out your story using different color Post-it notes on an arc (see Figure 7.5). Each color would be for a different user type. Or maybe you want to analyze a more complex set of data so that you can figure out what the story is. In this case, you might map out lots of data points as a narrative (see Figure 7.6).

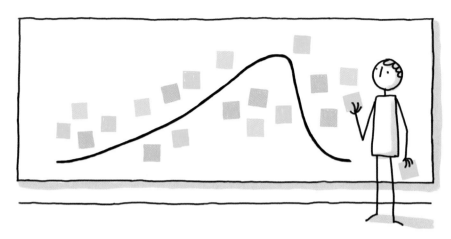

FIGURE 7.5

A behavior analysis for multiple user types.

FIGURE 7.6

A behavior analysis for multiple user types organized as a table.

- **Needs assessment:** What are all the things you need to support your story? Visualizing your story on a wall will help you map out your requirements. This can be simple, where you add Post-it notes as you see fit, or more complex with different rows for different things like *front-end* requirements and *back-end* requirements, or *haves* and *don't-haves* (see Figure 7.7).

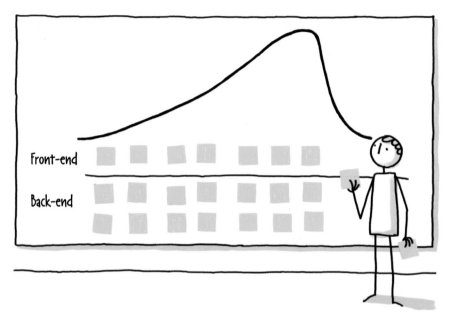

FIGURE 7.7
A needs assessment broken down by different categories.

- **SWOT (Strengths, Weaknesses, Opportunities, and Threats) analysis:** Storylines are a powerful, yet lightweight way to explore and visualize strengths, weaknesses, opportunities, and threats in a journey and more importantly in a product or service. Mapping your storylines out in this way will help you uncover not only what a story could be, but also what works, what doesn't work, what's missing, and what could be (see Figure 7.8).

FIGURE 7.8

A visual analysis of strengths, weaknesses, opportunities, and threats analysis. This can be used to assess a concept or a journey. Visually mapping the story in this way helps you see when you are missing things like a high point or a value and perhaps have many threats to consider.

Write Your Story

While visualizing stories is the first step toward making sure that you've got a solid, good story, writing that story out helps you uncover additional strengths, weaknesses, gaps, and opportunities that you might otherwise miss. Writing stories out with words is also the first step toward doing something else: communicating your stories, both internally with a team or stakeholders and externally with the world and your users.

Six-Word Stories

No one is sure where the format originated, but there is an origin story that goes something like this: Ernest Hemingway was hanging out with his literary friends. He claimed that he could tell a story in under 10 words. They didn't believe him. Bets were placed. In a bout of inspiration, Hemingway wrote these six words:

"For sale: Baby shoes, never worn."

Hemingway won.

While this story might not actually be true, it has inspired an entire literary subgenre of microtales or *six-word stories*. When you write a story with words, you don't need to spell out every single thing that happens. Sometimes, you only need a few words to allude to something, and you can let your reader fill in the blanks. In fact, sometimes less *is* more—more engaging, in this case, as your brain starts parsing meaning and completing the picture.

A short, minimally worded story, when carefully crafted, can act as a prototype for something much larger. More importantly, a short phrase or sentence can help you remember what your story is about as you find, explore, develop, and test it. Think of a short phrase in this case as your shorthand to use internally or externally.

For example, the book you're reading is a product. I've got a Post-it note on my wall reminding me what this book is about as I type. *Think like a storymaker.* I'm not a marketing professional, and that sentence might never see the light of day in any marketing capacity. But what it does is remind me of my goal, what I'm writing about, and keeps me on point. I want you to think like a story*maker* and make awesome stories for your users and business.

At FitCounter, we had a shorthand sentence for our concept story: *bite-sized, personalized fitness training, on your own time.* It wasn't marketing-friendly copy by any means. But it kept us on track. Our origin story was something like: *feel what it's like to get fit.* It wasn't pretty, but it also kept us on track.

Scenarios

While short punchy phrases and sentences are helpful to get and stay on track while, during, and after you explore and build your stories, writing longer scenarios gets a different part of your brain functioning. Writing scenarios works much like journey mapping, except that it's a writing exercise rather than a visual exercise. Scenarios help you use words to uncover how someone might interact with your product or service, as well as what will be required to support the story.

Use Cases and Agile User Stories

Use cases and, more specifically, Agile user stories are a tool that technology teams, especially Agile development teams, use to organize, make sense of, and phrase requirements. Outlining use cases and user stories helps give context to what you build, why, and how. For developers, these tools are important because, much like other humans, developers are story-based creatures. They want to know what the story is before they spend hours, weeks, months, or [gasp] years building something. While Agile user stories are an excellent tool and artifact to envision, articulate, and communicate a body of work with a team, they're even more powerful if they have a story arc at their foundation. This architecture can bolster an entire body of work, as well as the smaller stories that make up that body of work.

Agile user stories typically are written out in this format:

As a [type of user], I want to/can [do something] so that [result].

At FitCounter, we used to have user stories that read like this:

As a [visitor], I want to [sign up for an account] so that [I can watch videos].

And you know what? No one wants to sign up for an account so that they can watch videos. That's a terrible story. I have the data to prove it—both quantitative and qualitative.

After we finally figured out what all of our storylines were and could be, we eventually started crafting stories that sounded more like this:

As a [visitor], I want to [sign up for an account] so that [I can save my personalized fitness plan for future use].

What we eventually figured out was that people would sign up for an account if they got *value* (climax) rather than just got *access* to something (anticlimax).

While most of this story development happened before we wrote our requirements out as user stories, we made sure that the story was apparent so that developers weren't building without context. This not only helped the product team align and communicate goals to designers and developers, but it also helped us quickly and effectively assess when our story was missing or when we were building things for no good reason.

Act It Out

Once you've got an idea of what your stories and storylines can be, you can also act them out. Doing so helps you uncover *more* strengths, weaknesses, and opportunities than you do visually or with written words. Additionally, acting your stories out is a great way to prototype and test your stories with real-life humans. Stories need structure, yes. Stories also need to be human-friendly.

Improv and Role-Playing

A good interaction between a person and a product or service functions like a good conversation. As such, you can and should test this conversation out with your team (or even stakeholders or clients) to make sure that the flow is not only sound, but also moves your story forward.

At FitCounter, we live-action prototyped our initial proof of concept, as well as many of our flows and storylines, internally and with potential customers to make sure that we were on the right track.[3] We even reverse engineered stories from real-life conversations with physical trainers so that we understood how the product could work.

Here's how a real-life interaction with a trainer and our internal improvisation (improv) sessions played out: imagine you want to start training. You join a gym and get assigned a trainer. When you meet with your trainer for the first time, she might ask you a series of questions so that she can determine your goals and fitness level. At

---

3 For more on live-action improv, see Gray, Dave, Sunni Brown, and James Macanufo. *Gamestorming: A Playbook for Innovators, Rulebreakers, and Changemakers.* Sebastopol, Calif.: O'Reilly, 2010.

that time, she will make sure to hit on your pain points so that you really know why you're there (exposition).

Then she might affirm or adjust your goal and tell you that if you work with her, you can attain this goal (inciting incident). Next, you'll do a series of exercises (rising action) where your trainer might purposely test your strength, endurance, or agility to see how far you can go. These exercises should be easy enough to partially complete (more rising action), but difficult enough so that you either find them physically strenuous by the end or cannot complete them (crisis). Fear not, the trainer tells you. If you stick with her, within a few weeks, you'll be that much closer to running that 5K or easily doing 30 push-ups or whatever goal or obstacle you want to meet or overcome (climax). She might even show you a visualization that projects how well you will do if you stick with the plan and exercise a few times a week. At this point, you not only hear her words, but you also feel like you just did something and can see yourself coming back weekly to do this until you meet your big goal (falling action). The trainer gives you your 12-week plan, and you're on your way (end).

Improvising and re-creating stories like these are how we engineered everything from on-boarding to customer service scripts to payment flows. Conversations, role-playing, and improv sessions aren't just about two people talking and saying words aloud; they embody stories that resonate with people and get them that much closer to meeting a goal. If you try this kind of activity with your team, just remember to consider the story. Nothing is more boring for your customers than a conversation that doesn't go anywhere.

Elevator Pitch

An elevator pitch is a short statement, sentence, or a few sentences you use to briefly describe a concept, product, or business. It's called an "elevator pitch" because it should be short enough that you can give the pitch in the time it takes to ride in an elevator with someone. Elevator pitches are not just marketing or sales tools—they are a strategic tool you can use to make sure you are clear on a product or project's objectives, market, and value to an intended audience. If you break the format of an elevator pitch down and consider it within the framework of a narrative, it is essentially a very short story.

You can use an elevator pitch to communicate your story to investors, stakeholders, customers, or team members in a fast, portable way that storymaps can't help you do. If you're at a cocktail party, meeting with investors, or simply want to get an idea across quickly and effectively, you won't always be able to draw or show story arcs. But you can speak a few words and see how people respond. Just remember, your story architecture should flow through your elevator pitches so that they are that much stronger. Just like a short and sweet story, your elevator pitch must grab people's interest, take them on a little journey, and make them see the value in what you're talking about.

Elevator pitches for products and services come in different formats and generally function like this:

> For [target customer] who has [customer need], [product name] is a [market category] that [one key benefit]. Unlike [competition], the product [unique differentiator].[4]

If you reverse engineer the format, you can see how this maps out onto and functions much like a concept story:

- Target customer = main character

- Customer need = inciting incident or problem

- Product name, category, one key benefit = rising action

- Competition = crisis and conflict

- Unique differentiator = climax, resolution, or what's *awesome*

- And the falling action and end = this is how it all comes together: a customer should be able to use a product not only to get something done, but also to see the benefit and the value. Conflict overcome, crisis averted.

Putting It All Together

Once you've got a good idea of what your storylines can be, you'll want to continue to weave them into everything from the actual physical or digital prototypes that you build to the way you present your work to how you test your products with your target market. Stories aren't just something that should stay in your workplace or with your team—they eventually need to make it to the outside world. Weave

---

4 While there are many different types of elevator pitch formats, I like to use this one: http://www.gamestorming.com/games-for-design/elevator-pitch/

them through everything you do related to creating a product, including pitching, presenting, and demonstrating your ideas.

Build and Communicate

Consider how story ultimately wove through the first iPhone. When Steve Jobs gave his keynote presentation in 2007 announcing the iPhone, he had a vision for the product, and knew how to weave that vision into everything from the design, the presentation, and even the prototype of the first iPhone. It turns out that the product he demonstrated that day to illustrate his point that Apple was reimagining the way we communicate wasn't an actual functioning iPhone—it was a semi-functional prototype.[5] As such, the device was essentially a prop that supported the overall vision and story as he walked the world through a series of smaller storylines that introduced features and functionality like the ability to pinch and zoom into a high-resolution photo or search for Starbucks on a map, call the store, and order 1,000 lattes.

"And here we are...boom!"

Those are the words that Steve Jobs used during his presentation to narrate how you search for and locate something in the Google Maps app on the first iPhone as he demonstrated the device onstage. As behind-the-scenes accounts now tell us, the prototype was only marginally functional; it could barely catch a Wi-Fi signal, make a phone call, or operate without crashing. And as owners of the first iPhones remember, the phone was not at all fully featured compared to other smartphones on the market. The Palm Treo, which was the best-selling smartphone the year before, not only had features like cut-and-paste, but it also let you install apps and games. The iPhone did neither. But what the iPhone and in particular the cobbled-together prototype that Jobs demonstrated did do was go "boom." Just like a good story.

The perfectly timed "boom" during this demo and overall presentation was not a coincidence. In fact, it is just like that one plot point from the fourth season of *Breaking Bad*: *BOOM* (see Chapter 1, "Mapping the Story"). Only instead of actual explosions, the "boom" Jobs was narrating was a tiny animation that transpired when a pin

---

5 Vogelstein, Fred. "And Then Steve Said, 'Let There Be an iPhone'." *The New York Times*, 04 Oct. 2013.

dropped on a map (see Figure 7.9). This animation was so fast and tiny that if you weren't looking for it, you missed it. But if you spent years working at Pixar or being forced to watch and analyze scenes and sequences over and over, you noticed it. And if you spend some time after reading this book looking for and analyzing sequences and flows like the one for adding a calendar event in Fantastical, you'll start to notice them, too. Nothing is too small, fast, or insignificant to inject into a design or prototype as long as it supports the story. If the story is strong and compelling enough—like the ability to search anywhere on the globe using your phone, which was a revolutionary concept in 2007—you not only save that demo for last, but you also emphasize the story with narration. Just like Jobs did.

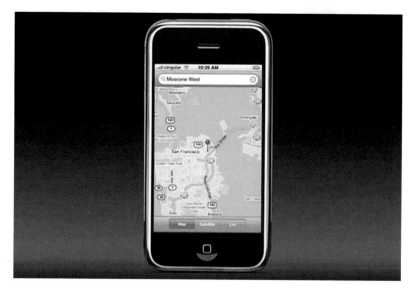

FIGURE 7.9

While the first iPhone didn't have basic functionality like cut-and-paste, the pin-dropping animation was meticulously crafted for bounce and elasticity.

Prototyping, demonstrating, and presenting your work, whether you are a designer, developer, product manager, executive, or founder is essential for conveying why your product is great *and* for explaining how it works and why it matters. Plus, it's essential for getting buy-in, whether it is from an internal sponsor team, clients, or the world audience.

At FitCounter, we wove our storylines into everything from research to design, to strategy and presentations, demonstrations, and even board and investor pitch decks. And it worked. Without a story, our customers and potential investors were confused. With a story, they were excited. The great thing was that both investors and customers put their money where their mouths were. So much so that the actual start-up that the story is based on is doing great—they have a story and the revenue generation to show for it.

Again, this is not just film magic or some kind of smoke-and-mirrors illusion or sleight of hand. It's good old story craft that is built for how humans interact with and understand the world around them. It works for the idea of a product, the actual product, and how you present that product to the world. It not only helps you sell your products, but it also makes them better, overall. More importantly, a story-first approach helps your customers have a better experience with that product. There is not enough magic in the world that can make humans unwittingly fake that on their end. Designing with story is designing for humans.

If you ever want to learn from the master of product stories, I recommend sitting down for an hour one day and watching that first iPhone keynote address from start to finish. In that presentation, you will see how someone uses story and story structure to build excitement about concepts and use cases. In the presentation, the iPhone isn't the star—rather, *you* are the star as you envision all of the things that you could do with this device in your hand. As an origin story, this story ends with you eventually buying the device as many people did and continue to do around the world. More importantly, however, it ends with you having a new and better way to do what you didn't realize you needed to do: communicate.

Was this presentation crafted with a clear overarching storyline and subplots? Absolutely. The best presentation experts in the world advocate using story architecture to bolster presentations. Was Steve Jobs aware of how powerful story structure could be when he demonstrated the smallest subplot or usage story for searching for and finding a location on a map? Undoubtedly. He spent years before going back to Apple working with Pixar—a bastion of visual storytelling and animation. As any filmmaker will tell you, everything you put in the scene must support the story.

While storytelling comes naturally to many of us, it is worthwhile to meticulously and carefully map out everything from how a prototype functions to how you ultimately present that prototype or a broader product idea to an audience. The better the story, the more engaging both your prototype and presentation will be.

Test and Validate

Stories are not just for building and communicating. They must resonate with your audience. Every step of the way, you must test your stories to make sure that you're on track. Diagrams, storyboards, written words, pitches, improv, prototypes, demos, presentations— these are all artifacts and activities you can use to test your stories with your target audience to make sure that you're on the right track.

How do you know you're on the right track? First, once you start to use story more in your work, you'll start to see when something is or isn't a solid story. Then, when you start putting your ideas in front of people, you'll start to get feedback. Maybe a story will or won't resonate with someone. Maybe people will tell you when something doesn't make sense to them. Maybe they'll smile at just the right time during a usability or concept test, validating your hunch that the climax was what you thought it would be. And if all goes well, maybe even journalists will hear about your product, use your product, and unwittingly echo your story.

When a story is structurally sound, it flows through *everything*. It's not only up to you to make sure that you engineer around the story, but also that the story resonates with and is echoed by the outside world.

Rules of Thumb

"The true writer is one for whom technique has become, as it is for the pianist, second nature."

—John Gardner,
The Art of Fiction

A few years ago, I stormed into a CEO's office and proudly proclaimed my sudden epiphany: "We have no story. We need a story!" I continued. "So...what's the story?" This was the start-up that I have fictionalized as FitCounter in this book. Luckily, the CEO still talks to me today. But at that moment, he turned bright red and asked me to leave his office. Stories are powerful tools. When used well, they make things better. But use them carefully. As I learned that day, no one likes being told that his product and something he's worked for years to build has no story. What people do like are products that are successful, customers who are happy, and metrics that help board members sleep at night.

Once you start working with story, you will weave it into your practice in many ways, shapes, and forms. You will wield story like a mighty sword as you break away and rebuild how you hope people will think about and experience your products. But remember— swords can hurt. What eventually worked for this CEO and most of my clients and teams ever since is not *telling*, but *showing*. If you're fixing something that's broken, use story not just to show what *is*, but also what can *be*. And if you're building something from scratch, use story to make some magic.

Here are some rules of thumb to guide you on your journey.

Stories Are Character-Driven

Remember, in the case of the stories you build, whether they are visual, verbal, analog, or digital, your main character must be the person on the other end, i.e., the person experiencing the story. Everything you want that person to experience must drive the story forward. If it doesn't, you cut it out. Also remember that the characters in your stories are not fictional or hypothetical dream customers who don't exist in real life. They are based on real people, real goals, real behaviors, and real stories. If you don't have access to real people or real data, you can make it up—as long as you validate it later.

Characters Are Goal-Driven

It is your job to drive your character's story forward. It is also your job to help that person meet his or her goal. There is nothing worse than a story with no purpose, whether it is for people in an audience watching a movie or for the people using your product. When crafting your story, make sure to move your character forward *and* to help him (and your business) meet some kind of goal.

Goals Can Change

Humans are fickle. In romantic comedies, the protagonist often starts out with a simple goal: to end up with the guy who is no good for her. Over time, she will probably learn that what she really needs to do is be with the nice guy. That said, what you *want* and what you *need* aren't always obvious in the moment and that's OK.

In 2006, people said that they wanted an iPod that made phone calls. In 2007, instead they got a new way to communicate. If you think you know what a character's goal is, consider asking "why" as many times as you can until you uncover the real goal at the core of your story. You want an iPod that makes phone calls? Why? Because it would be convenient. Why? Because you only want to carry one device. Why?...and so forth.

Also know that while goals can change, characters won't always know what their goal is until you uncover it for them. I have yet to meet someone who tells me that his goal is to communicate with the world around him. But I've worked on enough products where that is undoubtedly the main character's goal. Sometimes, it's your job to remind people and get them on the right track.

Goals Are Measurable

Imagining stories that help people meet their goals isn't simply an exercise in creativity. Goals in the stories you create can and should be measured, both for your characters and your business. You can do this by measuring qualitatively (in-person conversations and observational studies) and quantitatively (surveys and analytics)—or using a combination of the two. You can measure actions, and you can measure sentiment, ideally both. Doing so requires measuring what people think and do, both of which are essential for stories to move forward and help characters meet their goals. Observing stories in action and measuring goal completion is how you know you're on the right track.

Conflict Is Key

Once upon a time, there was a wooden doll named Pinocchio. He wanted to become a real human boy. So he went out, bought some magic, and transformed into a boy. The end.

Easy, right?

Of course not.

Stories wouldn't be stories without conflict. They would be a series of events—stuff happening. Just as stuff happening doesn't make for good stories, it definitely doesn't make for engaging experiences.

Every character and every goal in a story should always have some kind of force working in opposition to it. Conflict makes everything that happens in a story more suspenseful and ultimately rewarding, whether you experience a story by watching it from the outside or by participating in the story, as you do when you engage with products and services.

Designing stories for conflict helps you determine what should happen next as you strive for balance between a character trying to attain a goal (or set of goals) and running into obstacles along the way. When building products around stories, it is not your job to place obstacles in the way, but instead to consider them and plan around them as you help your protagonist jump hurdles, have a smooth ride, and ultimately meet his and your business goals.

Math Is Fun

Here's a formula you didn't learn in math class. I use it all the time to make things better.

$$A \times B = C$$

A represents forward momentum, which is the force that a character exerts to meet his or her goal with your product or service. *B* is the force or forces that act against your character and get in the way of that person using your product to meet his goal. If you multiply the forces, you get *C*—and that's what will ultimately help the climax or the way that the customer uses your product to resolve this conflict (see Figure 8.1).

FIGURE 8.1
Opposing forces collide to form a climax or a peak.

I use the iPhone as an example to illustrate this law because it's so easy to refer to when I'm in meetings or too tired to think (which is often). If Force A is that you want a 2-in-1 device so that you can communicate better. And Force B is that you don't want to buy a new device or are afraid that it will be difficult to use. Then C is what wins: a 3-in-1 device that works like magic and is easy to use. It's so simple to use that Steve Jobs ordered 4,000 lattes from Starbucks at the tap of a screen around 3/4 of the way through his 2007 keynote presentation. He didn't just *say* that it was a 3-in-1 device that worked like magic. He *showed* how 3-in1 *was* magic.

Choose Your Own Adventure

Storylines are linear. The products you build are not. They involve complex systems, decision points, branching paths, interactions, feedback loops, dependencies, and infinite permutations. One question I'm often asked is how and why you would use such a simple, linear framework like story to envision and plan interactive products and services that are anything but simple. Shouldn't we plan for something that resembles more of a *Choose Your Own Adventure* novel that you might have read when you were young? Or a modern corollary—video games?

While complex products have much in common with complex entertainment media like novelty books and video games, defining complexity as branching rather than linear is taking a system-centric approach. If you want to engage someone using your product, you want to design for the human experience, not the system. Systems are complex. Human experiences are linear. That's because experiences happen as a series of moments in time. Until flux capacitors and time-shifting Delorians are a reality, time is unfortunately linear.

Whether someone using your product chooses to sign up now or sign up later, visit this page or that page, perform this task or that task, and do so successfully or hit a roadblock or error, the constant for that person is time. As such, when using story to assess, envision, and plan for intended experiences of use, embrace complexity by thinking linearly, one story at a time. Each character, each use case, each scenario, the happy paths, the critical edge cases, they each get their own story. Different stories might intersect and have common features and plot points (see Figure 8.2). And seeing the intersections and commonalties can be valuable. But each story lives on its own and should be given due diligence. Something like a flow chart is wonderful for mapping out a system from a birds-eye view. Stories are a map of the human experience, from the human view. Yes, in reality, your users will choose their own adventures. And yes, you can and should plan ahead so that those adventures are as engaging as can be.

FIGURE 8.2
Different storymaps comingle in an ecosystem of stories, each unique to a different character user type.

Make Things Go *Boom!*

Story structure works the way it does because humans need a little something near the end of an experience to help them pay attention, remember, learn from, and see value in what they just sat through or did. Or maybe humans evolved to need this something because of millennia of communicating through story.

No matter why, the fact remains: If you want to have maximum impact and help people and businesses meet their goals, make things go *boom*!

This *boom* can be as seemingly insignificant as a little animation on a tiny screen or as big as a voucher for a free flight because someone sat on the tarmac on your company's airplane for three hours and complained to your customer service team via Twitter. As long as that little something is significant, delightful, or impactful, the experience will be memorable and the story repeatable…maybe even never-ending.

INDEX

ACKNOWLEDGMENTS

For all of my students, workshop attendees, and clients who asked me for a book to read so they could hone their narrative craft for product design. There was no such book. For you, now there is.

For those who helped me start writing and counseled me through the way: Lou Rosenfeld, Kevin Hoffman, Karen McGrane, Christina Wodtke, Abby Covert, Sara Wachter-Boettcher, Rebekah Cancino, Matt Grocki, Tomer Sharon, Jeff Gothelf, Jonathon Colman, Russ Unger, and Dave Gray.

For those who read early drafts. You pushed me to do what I hate most and use my words to explain myself: Lis Hubert, Margot Bloomstein, Jane Pirone, Maya Bruck, Jonathan Berger, and Bill Gullo.

For those who contributed insight, ideas, and imagery along the way: Marta Justak, Eva-Lotta Lamm, Ajay Rajani, Lis Hubert, Michael Leis, David Malouf, Chris Noessel, Paul Rissen, and Senongo Akpem.

For my fellow campers. You know who you are. You remind me daily that it's OK to take up space.

And for Snowball, my cat, who passed away right before I decided to write a book. You taught me that a cardboard box is never just a box. It's whatever you want it to be.

ABOUT THE AUTHOR

For more than 15 years, **Donna Lichaw** has guided startups, non-profits, and Fortune 500 brands in optimizing their digital products and services by providing them with a simplified way to drive user engagement through impactful storytelling. As a consultant, speaker, writer, and educator, she utilizes a "story first" approach to help teams define their value proposition, transform their thinking, and better engage with their core customers.

Donna developed her talent for storytelling and narrative development as a documentary filmmaker, from which she built a successful career as a digital product strategist for a number of emerging and established companies in New York and London. She has been recognized as a subject expert on storytelling and customer engagement strategies, speaking at design and technology conferences throughout the U.S., Canada, and Europe.

Donna graduated from Northwestern University with an MFA in radio, film, and television, and completed her undergraduate degree in film and video studies at the University of Michigan in Ann Arbor. She has taught courses and delivered seminars on design, communication, and user experience at New York University, Northwestern University, and General Assembly, and is an adjunct faculty member at the School of Visual Arts.

She currently works and resides in Brooklyn, New York with her partner, Erica, their crooked dog, Ralph, and doe-eyed, pointy-eared cat, Gizmo. You can find her at **www.donnalichaw.com** and on Twitter **@dlichaw**.